Microbial quality assurance in cosmetics, toiletries and non-sterile pharmaceuticals

LEADING EDGE BOOKS IN PHARMACEUTICAL SCIENCES

**Books are to be returned on or before
the last date below.**

0008 ·

2 0 FEB 1997

2 2 FEB 2000

0 2 MAY 2003

0 7 MAY 1997

- 2 MAR 2000

2 3 MAY 1997

1 0 MAR 2006

2 6 JAN 1998   1 6 MAY 2000

2 8 FEB 2001

2 3 MAR 1998   0 5 APR 2001

1 7 OCT 2001

Microbial quality assurance in cosmetics, toiletries and non-sterile pharmaceuticals

Second Edition

Edited by

R. M. BAIRD

formerly NETRHA Pharmaceutical Microbiology Laboratory, St. Bartholomew's Hospital, London

with

S. F. BLOOMFIELD

Chelsea Department of Pharmacy, King's College London (KQC)

Taylor & Francis
Publishers since 1798

UK	Taylor & Francis Ltd, 1 Gunpowder Square, London, EC4A 3DE
USA	Taylor & Francis Inc., 1900 Frost Road, Suite 101, Bristol, PA 19007

British Library Cataloguing in Publication Data

A catalogue record for this book is available from the British Library.
ISBN 0-7484-0437-6
(formerly 0-13-101247-9)

Library of Congress Cataloging-in-Publication Data are available

Cover design by Jim Wilkie, Emsworth, Hampshire.

Typeset in Times 10/12pt by Heather FitzGibbon, Christchurch, Dorset.
Printed in Great Britain by T. J. Press (Padstow) Ltd.

Contents

Preface

As a subject, microbial quality assurance continues even today to tax the minds of those involved in the manufacture of cosmetic, toiletry and non-sterile pharmaceutical products. Over the years scientists working in this field have developed increasingly sophisticated measures to minimize the risk of product contamination. By not only understanding the problems of microbial contamination, but also appreciating their significance, considerable improvements in contamination control have been made through implementation of good manufacturing practices and the application of new technologies. Despite these many advances, problems of microbial contamination continue periodically to cause concern, albeit less frequently than in the past.

The first edition of this book was published in the wake of a meeting held at London University at a time of considerable change in working practices. In combining information from many scattered sources, it presented a unique review of the state-of-the-art approaches not only to contamination control, but also to the broader issues of microbial quality assurance in general. The book was consulted by a wide readership and served as a valuable reference source for several years.

In compiling the second edition, the opportunity arose not only to produce a 'stand-alone' text with both new and substantially updated information, but also to re-organize the presentation of material into four sections, thus guiding the reader through a step-wise approach to contamination control. At the same time the editors have sought to provide an authoritative overview of the microbiological issues affecting the manufacture and quality assurance of cosmetics, toiletries and non-sterile pharmaceuticals, compiled into a single book. Thus guidance is given on wide-ranging microbiological issues, extending from early formulation considerations and manufacturing requirements for different dosage forms to the testing of the final product against internationally agreed standards and guidelines. In this edition it will be seen that particular emphasis has been placed on issues affecting both toiletry and cosmetic products, thereby addressing many of the current consumer concerns. The appearance of this edition is both timely and appropriate and it is hoped that a fresh complexion is presented by the new format.

Dr Rosamund Baird
Yeovil, UK

PART ONE

Identifying the issues

Control of microbial contamination in cosmetics, toiletries and non-sterile pharmaceuticals

SALLY F. BLOOMFIELD

Chelsea Department of Pharmacy, King's College London, University of London, Manresa Road, London SW3 6LX

In the manufacture of cosmetics, toiletries and all types of pharmaceuticals, quality assurance represents a major consideration. This book is particularly concerned with one aspect of quality, namely the assurance that products are not contaminated with organisms which might affect their safety, efficacy or acceptability to the consumer or patient.

One obvious approach to this problem is to ensure that all products are manufactured either as sterile products or as 'single-use' packs. For cosmetics and toiletries and many pharmaceuticals, this is neither appropriate nor commercially viable and alternative means of microbial quality assurance are therefore sought. During product manufacture, microbial contamination is mainly controlled by the application of good manufacturing practice (GMP) whilst the maintenance of quality during storage and use (also the responsibility of the manufacturer) is achieved largely by the inclusion of preservatives but also by product design and by product formulation and packaging.

In practice, the presence of micro-organisms in pharmaceuticals, cosmetics and toiletries constitutes a potential hazard for two reasons. These aspects are reviewed in chapter 2. Firstly, it may result in spoilage of the product – the metabolic versatility of micro-organisms is such that almost any formulation ingredient may undergo degradation in the presence of a suitable micro-organism. Alternatively, it may constitute an infection hazard to the consumer or patient, although here we have to bear in mind that the degree of hazard will vary considerably from one situation to another according to the intended use of the product (oral, topical, application to the eye, etc.) by the patient or consumer.

In attempting to define acceptable standards or limits for microbial contamination for non-sterile products, it is therefore extremely difficult to establish what levels and types of contamination represent a hazard and what can be considered as safe.

If the problem is examined from a historical point of view it is found that, prior to about 1960, although products were manufactured under hygienic conditions with the inclusion of a preservative, there was little evidence of the rigorous approach which is currently adopted. During the 1960s, following increasing numbers of reports of infection outbreaks associated with contaminated products (see chapter 2), there was a general realization of the need for improvements in microbial quality assurance for all types of products.

In order to assess how this might be achieved, a number of studies were initiated to obtain a general picture of the quality of products being manufactured at that time. Some of the studies were official or national studies as in Sweden and Denmark (Kallings *et al.* 1966; Ulrich, 1968), whilst others were independent surveys. The investigations included cosmetics and toiletries as well as pharmaceuticals. In the UK, two official studies of pharmaceuticals were initiated, one by the Public Health Laboratory Service to study hospital medicaments, the other by the Royal Pharmaceutical Society of Great Britain (RPSGB) to study all types of products (Anon. 1971a, 1971b). The results of all these surveys are reviewed in chapter 15.

Having established the nature and extent of the problem, the next step was to try to control it; the various official bodies had to decide the extent to which manufacturers should be asked to limit microbial contamination in their produces and how the standards should be applied. Inevitably, the limits varied from one country and one situation to another. Accepted limits may be 'in-house' limits or, for pharmaceuticals, may be applied via the pharmacopoeia or licensing systems. The various official, unofficial and 'in-house' standards which have been developed over the last 20 or so years are outlined in chapter 15.

In examining the methods which are currently adopted for controlling microbial contamination in pharmaceuticals, cosmetics and toiletries, it can be seen that in general these have been developed from specific recommendations laid out in the original RPSGB working party report and the report of a joint working party representing the Toilet Preparations Federation and the Society of Cosmetic Chemists (Anon. 1970). The recommendations of the RPSGB were then further developed and incorporated into the UK *Guide to Good Manufacturing Practice* (Anon. 1983). As a result of European harmonization, the UK guide has been superseded by the European Guide to Good Manufacturing Practice (Anon. 1992, 1993a). Recommendations relating to the manufacture of cosmetics and toiletries are outlined in a document published by the Cosmetic and Toiletry Products Association (Anon. 1993b).

If we refer to these recommendations, we find that they divide into three groups: recommendations related to GMP, recommendations related to formulation of the product, and recommendations which related to the inclusion of preservatives. In the following chapters of this book, each of these aspects is considered in detail. The problems of applying these various principles to microbial quality assurance of two specific groups of products, namely powder, tablet and capsule formulations, and hair and skin products is also discussed in chapters 7 and 8 respectively.

As far as GMP is concerned, the main concerns are with the microbial quality of the raw materials (particularly water) and with all aspects of the manufacturing process, the processing environment and equipment, the process itself and the personnel operating the process. These aspects are discussed in chapters 3 and 4.

As far as product formulation is concerned, any number of physicochemical factors can affect the fate of micro-organisms which enter a product. These factors include the conditions of water activity, pH and osmotic pressure within the product, the availability of nutrient material, the product storage temperature, and so on. In practice, it is found that some types of products are virtually 'self-preserving' such that residual contamination occurring during manufacture (or use) is reduced to undetectably low levels during storage. In the pharmaceutical industry, there has been some tendency to overlook this aspect and to rely almost entirely on chemical preservatives. With increasing concern over possibilities of toxic side effects associated with chemical preservatives, it is important to consider carefully the possibilities of 'self-preservation or natural preservation' of products. This aspect is discussed in chapter 5.

The third aspect relates to the inclusion of chemical preservatives. Although the microbial quality of products should be achieved by GMP, it may be necessary in certain types of product to include a preservative to protect the product against residual contamination introduced during manufacture and any further microbial contamination which might occur during use. It is clearly understood, however, that the function of the preservative must not be to cover bad manufacturing practice but to ensure that the product remains in a satisfactory condition during storage and use.

In solving problems related to chemical preservation, there are four main questions which have to be considered. Under what conditions is the inclusion of a preservative justified? What constitutes adequate preservation? What preservatives are available to meet our requirements? How do we establish that a product is effectively preserved?

In deciding under what conditions the inclusion of a preservative is justified, for some situations the indications are obvious whilst in others there may be room for disagreement. Thus, for example, in tablet film coating where working schedules demand advance preparation of film coating solutions, although the inclusion of preservatives in these solutions might be considered as an admission of bad manufacturing practice, in practice this is considered acceptable to prevent microbial growth which would otherwise occur during the unavoidable holding period prior to the film coating process. In addition to this, although tablets are assumed to be self-preserving, the question has been raised as to whether preservatives should be included to prevent incidents of mould growth which occur from time to time. Problems associated with control of microbial contamination in powders, tablets and capsules are discussed in chapter 8.

Secondly, it is necessary to consider what constitutes adequate preservation. As far as pharmaceuticals are concerned, it appears that the licensing authorities currently allow themselves to be guided by criteria defined in the *British Pharmacopoeia (BP), European Pharmacopoeia (EP)* and *United States Pharmacopoeia (USP)* tests for preservative efficacy (Anon. 1993c, 1993d, 1995). If one asks how such standards have arisen, it would be difficult to get a precise answer. It would be comforting to believe that they represent a definitive knowledge of the required activity to protect the product and user against the appropriate hazard, but in practice, as mentioned earlier, such information is not available, nor is it ever likely to be. In effect, the criteria appear to be based on a knowledge of what is achievable in relation to the range of available preservatives and a working knowledge of preservative efficacy in manufactured products. In recognition of the problems of deciding what constitutes adequate preservation, and the fact that this may vary from product to product according to its nature and intended use, the *BP* test states that, at present, compliance with the test is not rigidly demanded; deviation is acceptable where it is adequately justified by bacteriological and other data (Anon. 1993c). For cosmetics and toiletries, at the present time there are no official standards, the industry preferring to work to 'in-house' standards which relate to levels of preservation demonstrated by established products known to be 'well-preserved' (Anon. 1993b). These aspects are further discussed in chapter 12.

Having considered what might constitute 'adequate preservation', the next problem is to find suitable preservatives which meet these requirements. Over the years, a very large number of antibacterial compounds have been examined and their suitability for preservation of pharmaceuticals, toiletries and cosmetics investigated. In practice, only a relatively limited range of about 13 preservatives is currently used in the majority of

Table 1.1 Antimicrobial agents currently used in pharmaceutical products

(1)	*Single-dose and multiple-dose injections*	
	Chlorocresol (0.1%)	Benzylalcohol (1.0%)
	Cresol (0.3%)	Phenylmercuric nitrate, acetate or
	Phenol (0.5%)	borate (0.002%)
	Chlorbutol (0.5%)	
(2)	*Eye drops and contact lens solutions*	
	Chlorhexidine acetate or gluconate (0.1%)	Thiomersal (0.1%)
	Benzalkonium chloride (0.3%)	Chlorbutol (0.5%)
(3)	*Oral liquids*	
	Methyl, ethyl and propyl *p*-hydroxybenzoate	
	(parabenz) (0.3%)	
	Benzoic and sorbic acid (0.3–0.5%)	
	Chloroform (0.25%)	
	Bronopol (0.5%)	
(4)	*Creams*	
	Parabenz (0.3%)	Cetyltrimethylammonium bromide
	Chlorocresol (0.1%)	(1.0%)
	Bronopol (0.5%)	Phenylmercuric nitrate (0.01%)

pharmaceutical products (see Table 1.1). For cosmetics and toiletries the list is somewhat longer; the most recent EEC directive (Anon. 1993e) indicates some 45 preservatives which are considered 'acceptable' and a further nine which are 'provisionally allowed'. The latest FDA survey (Anon. 1993f) indicates that the most widely used compounds in cosmetics and toiletries are the *p*-hydroxybenzoates, imidazolidinyl urea, isothiazolones and phenoxyethanol. Despite the much larger list of compounds potentially available for use in cosmetics and toiletries, a 1982 survey of some 20,000 formulations in the USA suggested that 97% of them contained only seven preservative types (Decker and Wenninger 1982). More than anything it is the problems of toxicity which limit the range of preservatives available for use in pharmaceuticals, cosmetics and toiletries. Problems of preservative toxicity and toxicity testing are discussed in chapter 11. The other main problem is that of cost. Although there is an urgent need to increase the range of preservatives suitable for products, the commercial advantages of providing new preservatives are generally outweighed by the substantial research and development investment required for their development. For this reason, the major research effort in this area is being directed towards better use of 'existing' preservatives including the use of potentiators and of synergistic preservative combinations. These aspects are covered in chapters 9 and 10.

In choosing a suitable preservative for a particular formulation, the compound is required to have certain characteristics.

(1) It should be active against the required spectrum of organisms.

(2) It must be effective at the pH of the formulation.

(3) It must be soluble in the aqueous phase of the formulation.

(4) It must be stable and non-volatile

(5) It must be compatible with formulation ingredients.

(6) It should be colourless, odourless and tasteless.

(7) It must be of low toxicity.

In practice, the ideal preservative does not exist, and in many cases it is a matter of selecting the 'least unsuitable' rather than the 'optimum' preservative for a particular product. For information relating to the properties of individual agents or groups of agent, and the selection of preservatives for individual products, reviews by Bean (1972), Hugo and Russell (1992), Cowen and Steiger (1977), Croshaw (1977), Bloomfield and Leak (1982), Block (1983), Kabara (1984), Geis (1988), Denyer and Wallhaeusser (1990), Orth (1993) and Anon. (1993b) should be consulted.

In addition to choosing a suitable preservative, or preservatives, for inclusion in a particular product, we also have the problem of maintaining an adequate preservative concentration within the product to ensure satisfactory preservation. Although simple aqueous formulations are relatively easy to preserve, many pharmaceutical and cosmetic formulations are complex systems consisting of a number of phases. Since microorganisms will grow only in the aqueous phase of a product, it is vitally important to maintain an adequate concentration of free preservative in the aqueous phase of the product. Problems can also arise from partitioning of the preservative into packaging materials. These aspects will be considered in chapter 6.

Finally, there is the problem of establishing that a product is effectively preserved. Although a knowledge of the properties of a preservative may assist in the choice of a suitable system, it is well recognized that selection based on theoretical considerations can be regarded only as a guide, since in many cases the interactions are incompletely understood. For this reason, it is always necessary to undertake microbiological tests to determine the efficacy of the preservative system for each individual product. This usually involves the performance of simple challenge tests in which a sample of the product is inoculated with the test organisms and the number of surviving organisms determined at intervals by viable counting procedures. The test system currently adopted for pharmaceuticals is usually based on the *BP*, *EP* and *USP* tests (Anon. 1993c, 1993d, 1995). Although these tests may be suitable for establishing the 'bioavailability' of the preservative within the product over the designated period of storage, under defined conditions, concern has been expressed regarding the adequacy of such tests for predicting preservative failures in the practical situation. These aspects and the various alternative approaches to preservative testing of pharmaceuticals, cosmetics and toiletries are discussed in chapters 12 and 13. New methodology for preservative testing is discussed in chapter 14, as well as the traditional and more recent rapid methods used in MQA testing of these products.

In the following chapters of this book, the individual aspects of microbial quality assurance as outlined in this chapter are discussed in more detail. For each aspect, an attempted has been made to review theoretical as well as practical information available and to illustrate how this information may be developed and applied in achieving satisfactory microbial quality assurance for cosmetics, toiletries and non-sterile pharmaceuticals.

References

ANON. (1970) The hygienic manufacture and preservation of toiletries and cosmetics. *J. Soc. Cosmet. Chem.* **21** 719–800.

(1971a) Microbial contamination in pharmaceuticals for oral and topical use: Society's working party report. *Pharm. J.* **207** 400–402.

(1971b) Microbial contamination of medicines administered to hospital patients: report by Public Health Laboratory Service Working Party, *Pharm. J.* **207** 96–99.

(1983) Guide to good manufacturing practice. HMSO, London .

(1992) Good manufacturing practice for medicinal products. In: *The rules governing products in the European Community*, Vol IV. Commission of the European Communities.

(1993a) *Rules and guidance for pharmaceutical manufacturers, 1993*. HMSO, London

(1993b) *CTPA guidelines for effective preservation*. Cosmetic and Toiletry Products Association, London.

(1993c) *British Pharmacopoeia, Addendum*. HMSO, London, Appendix XVIC, A191–A192.

(1993d) Efficacy of antimicrobial preservatives. In: *European Pharmacopoeia VIII* 14 Fasicule 16.

(1993e) *Off. J. Eur. Communities*, No. L203/24.

(1993f) Preservative frequency of use. *Cosmetics and Toiletries* **108** 47–48.

(1995) Antimicrobial preservatives – effectiveness. *United States Pharmacopoeia XXIII*. US Pharmacopoeial Convention, Rockville, MD, p. 1681.

BEAN, H. S. (1972) Preservatives for pharmaceuticals. *J. Soc. Cosmet. Chem.* **23** 703–720.

BLOCK, S. S. (1983) *Disinfection, sterilisation and preservation*, 3rd edn. Lea and Febiger, Philadelphia, PA.

BLOOMFIELD, S. F., and LEAK, R. E. (1982) Preservation. In: Wilkinson, J. B., and Moore, R. J. (eds), *Harry's Cosmeticology*, 7th edn. George Goodwin, London, pp. 575–606.

COWEN, R. A., and STEIGER, B. (1977). Why a preservative system must be tailored to a specific product. *Cosmet. Toilet.* **92** 15–20.

CROSHAW, B. (1977) Preservatives for cosmetics and toiletries. *J. Soc. Cosmet. Chem.* **28** 1–14.

DECKER, R. L., and WENNINGER, J. A. (1982) Frequency of preservative use in cosmetic formulations as disclosed to the Food and Drug Administration – 1982 update. *Cosmet. Toilet.* **97** 57–59.

DENYER, S. P., and WALLHAEUSSER (1990) Antimicrobial preservatives and their properties. In: Denyer, S. P. and Baird, R. M. (eds), *Handbook of microbiological quality control in pharmaceuticals*. Ellis Horwood, Chichester, pp. 251–273.

GEIS, P. A. (1988) Preservation of cosmetics and consumer products: rationale and applications developments. *Dev. Ind. Microb.* **29** 305–315.

HUGO, W. B., and RUSSELL, A. D. (1992) Types of antimicrobial agents. In: Russell, A. D., Hugo, W. B., and Ayliffe, G. A. J. (eds), *Principles and practice of disinfection, preservation and sterilisation*, 2nd edn. Blackwell Scientific, Oxford, pp. 89–113.

KABARA, J. J. (1984). *Cosmetic and drug preservation. Principles and practice* (Cosmetic Science and Technology, Series 1). Marcel Dekker, New York.

KALLINGS, L. O., RINGERTZ, O., SILVERSTOLPE, L., and ERNERFELDT, F. (1966) Microbiological contamination of medical preparations. *Acta Pharm. Suec.* **3** 219–228.

ORTH, D. S. (1993) In: *Handbook of cosmetic microbiology*. Marcel Dekker, New York.

ULRICH, K. (1968) Microbial content in non-sterile pharmaceuticals. *Dansk Tidsskr. Farm.* **42** 1–49, 50–70, 71–83, 125–131, 257–263.

2

Hazards associated with the microbiological contamination of cosmetics, toiletries and non-sterile pharmaceuticals

D. F. SPOONER

formerly of Boots Company Limited, Radcliffe-on-Trent, Nottinghamshire NG 12 2 GZ

Summary

'Wise men learn by other men's mistakes; fools, by their own.'

Bacon

(1) Serious attention was not given to the hazards associated with contaminated non-sterile products until the 1960s. Patients, many with compromised defences, have been found to be at risk from organisms, particularly Gram-negative opportunistic bacteria, in non-sterile pharmaceuticals administered by different routes.

(2) Non-sterile preparations have acted as intermediate reservoirs in common-source outbreaks of hospital infection.

(3) Infection from contaminated cosmetics and toiletries does not appear to occur readily, but isolated incidents in hospitalized patients, and some eye infections, have been reported.

(4) The spoilage of non-sterile products may be recognized organoleptically but changes indetectable to the senses could be hazardous.

(5) Continuous vigilance is necessary by those responsible for the production, distribution and administration of non-sterile products in order to protect patients and users from these microbiological hazards.

2.1 Introduction

The possibility that infection arises from microbial contamination of parenterally administered preparations has been appreciated from the close of the last century. Many incidents were reported (Groves 1973, Holmes and Allwood 1979) and steps were taken to minimize their reoccurrence. Hazards arising from contaminated non-sterile products have been less readily appreciated. In the first half of the century, scattered reports of incidents of spoilage and infection appeared, particularly in connection with ophthalmic preparations, which were then generally non-sterile, and disinfectants and antiseptics. However, serious attention was only given to non-sterile products after reports from Sweden indicated that they could constitute a real and direct hazard to the patient (Kallings *et al.* 1966). The thorough Swedish investigations initiated many surveys of non-sterile products for microbial contamination in a number of other countries over the next decade. Such results were summarized by Sykes (1971), who mentioned two important investigations in the UK organized by the Pharmaceutical Society of Great Britain (Anon. 1971a) and the Public Health Laboratory Service (Anon, 1971b). Many preparations were found to contain a surprisingly high number of viable organisms, including some harmful species. Later, interest spread to cosmetics and toiletries and these products were also found to constitute a potential hazard by virtue of their bioburden (Baird 1977, Malcolm 1976). More recently, an extensive independent survey of the microbiological quality of cosmetics and toiletries at the point of sale in Europe was carried out. About 5% of nearly 5,000 samples were found to have total viable counts greater than 1,000 colony-forming units per gram (cfu g^{-1}) (some even exceeding 10^6 cfu g^{-1}), and a number of pathogens were isolated (Greenwood and Hooper 1982).

It has therefore been clearly established that significant microbial contamination of non-sterile products can occur. This constitutes a *potential* hazard, particularly to those people whose host defences are compromised, but it is the purpose of this review to indicate the circumstances in which such products have acted as a *real* hazard and have caused (or have been involved in) adverse reactions owing to the micro-organisms which they have harboured. Harm may result from the effect of spoilage, the initiation of infection directly or the facilitation of its spread. Most available information has originated from hospital practice where there is a greater chance of recognition of the connection between adverse reactions and contaminated products. Even in hospitals, such a relationship is not always obvious but chance has favoured the prepared mind and this has led to some fruitful investigations. In general practice the possibility of identifying the association is much more difficult, and published reports may therefore represent a minority of unfortunate events that have occurred. Certainly, reports involving contaminated cosmetics and toiletries are very sparse and, while it is possible that such events still go unrecognized, it seems probable that the healthy user has proved relatively unsusceptible to infections arising from this source.

Fortunately, man learns from his mistakes, and it is in this spirit that this summary of mishaps has been undertaken. A number of other reviews of different aspects of the topic have already appeared (Butler 1968, Parker 1972, Smart and Spooner 1972, Beveridge 1975, 1983, Somerville and Summers 1979, Baird 1981, Ringertz and Ringertz 1982, Bloomfield 1990). This chapter attempts to complement rather than to duplicate them.

2.2   Infection from non-sterile products

The most obvious and severe consequence of microbial contamination is the direct infection of the patient. Given that viable organisms exist in the preparation, the chance that infection results depends upon a considerable number of variables, as outlined by Ringertz and Ringertz (1982) and Parker (1984). The type of organism and the numbers present in the dose, or application, are probably paramount. The question of pathogenicity is a vexed one; probably any species of micro-organism may initiate an infection if conditions are suitable and, as virulence appears to be a fleeting characteristic, the factors involved are still poorly understood. Bruch (1972) pointed out the difficulty of deciding whether a micro-organism is objectionable in a non-sterile pharmaceutical. It is a complicated value judgement, which is particularly difficult for topical preparations. He contemplated the roles of many organisms in such preparations and concluded that most are objectionable for products applied to damaged epithelium, while others may be 'usually objectionable', depending on the species, the site and the health of the recipient. This concept of 'frank' and 'opportunistic' pathogens is fraught with difficulty because we still know so little about the factors affecting the properties by which a micro-organism causes death or reaction of the tissues invaded. Even absolute virulence, the ability of the organism to transcend all fluctuations of host resistance, is dependent on the number of organisms present. This has been established experimentally in normal volunteers and from chance observation of natural infection where, because the vector is known, the infective dose can be established. For example about 10^6 *Salmonella anatum* or *S. meleagridis* cells were found necessary to produce clinical symptoms of infection in healthy volunteers (McCullough and Eisele 1951) whereas, on average, only 50 or so cells of *S. napoli* constituted an infective dose for patients in an outbreak of salmonellosis following consumption of contaminated chocolate bars (Greenwood and Hooper 1983). Studies in volunteers have shown that the number of *Staphylococcus aureus* cells necessary to initiate a localized skin infection varies from less than 10^2 to more than 10^6, depending on local conditions (Marples 1976). In contrast, epidemiological evidence, coupled with extrapolation from clinical medicine and experimental pathology, has allowed tentative determination of the number of organisms necessary to initiate inhalation infections such as tuberculosis and anthrax (Williams 1967). Thus, infective doses vary considerably, even between similar species, but it is often difficult to decide whether this is a function of the organisms or of the host. Infection route is very important; for instance *S. aureus* and the anthrax bacillus are both far more virulent by the skin than by the alimentary tract. In contrast, many species, including *Salmonella*, infect via the mouth but cannot penetrate lightly abraded skin. Trauma obviously facilitates invasion of the skin by certain micro-organisms. This has been shown experimentally and hydration, superhydration, selective antimicrobial agents and a topical steroid have also been found to encourage experimental skin infections in volunteers (Marples 1976).

A number of natural conditions diminish host resistance and are particularly relevant to a consideration of infection arising from contaminated non-sterile pharmaceuticals. Age, genetic constitution, malignancies, diabetes and urinary tract abnormalities are all important while therapy with corticosteroids, chemotherapy, antimicrobial agents and radiation can facilitate infection. Since our knowledge of the hazards of microbial contamination is based to a large extent upon experience with non-sterile pharmaceuticals, these will be considered first in some detail.

2.2.1 *Non-sterile pharmaceuticals*

A number of reports of contaminated products, in some cases leading to infection, appeared in the 1960s. Before then publications were scant (Savin 1967). Sykes (1971) attributed this to the introduction in the 1960s of newer types of medicaments having more susceptible formulations and an appreciation of the need for better hygienic control.

11

It soon became apparent that no route of administration was entirely free of danger, as the examples given below, and summarized in Table 2.1 indicate.

2.2.1.1 Ophthalmic preparations

The eye is particularly vulnerable to infection, and ophthalmic preparations are now required to be sterile. However, this has not always been so and a number of unfortunate incidents occurred in earlier times (Savin 1967). Particularly notable were the loss of eyes after the use, during intraocular operations, of saline contaminated with *Pseudomonas aeruginosa* (Ayliffe *et al.* 1966) and eight cases of severe eye disorders caused by use of a cortisone ointment contaminated with the same organism (Kallings *et al.* 1966).

Contact lens solutions, used to clean, disinfect or wet the lenses, are now sterile products controlled in the UK by the Medicines Act (Anon. 1968). The lens constitutes a foreign body to the eye and it is not surprising that bacterial and fungal corneal ulcers have been observed in users (Templeton *et al.* 1982). Contact lens solutions must obviously be capable of dealing with adventitious microbial contamination and therefore contain preservatives. Their formulation in this regard is rather more demanding than for other multiple-dose sterile products because the lens can take up some preservatives from the solution and eye irritation may subsequently be caused after prolonged contact (Norton *et al.* 1974, Richardson *et al.* 1979).

Cases of keratitis have been traced to eye-droppers contaminated with *Serratia marcescens*. The organisms were isolated from the caps and appeared to have grown in this situation. As the eye-drop fluid was expressed into the eye via the dropper it was inoculated with *Serratia* (Templeton *et al.* 1982).

2.2.1.2 Oral administration

Infection with salmonellae has occurred on a number of occasions following administration of pharmaceutical products contaminated with this oral pathogen. The report by Kallings *et al.* (1966), from the State Bacteriology Laboratory in Stockholm, created considerable interest at the time. Thyroid tablets contaminated with *S. muenchen* and

Table 2.1 Non-sterile products as independent reservoirs of infection

Preparation	Infecting organism
Eye ointment	*Pseudomonas aeruginosa*
Eye-dropper	*Serratia marcescens*
Thyroid tablets	*Salmonella muenchen*
Carmine capsules	*Salmonella cubana*
Pancreatin powder	*Salmonella agona*
Lignocaine jelly	*Pseudomonas aeruginosa*
Inhalation medication	*Klebsiella* sp., *Serratia* sp., *Pseudomonas* sp.
Topical anaesthetic	*Pseudomonas cepacia*
Handcream	*Klebsiella* sp., *Pseudomonas aeruginosa*
Detergent	*Pseudomonas aeruginosa*
Talcum powder	*Clostridium tetani*
Cellulose wadding	*Pseudomonas aeruginosa*
Antiseptic	*Pseudomonas* spp., *Serratia* sp.

S. bareilly caused salmonellosis in more than 200 people throughout Sweden. Since *S. muenchen* infections had been comparatively scarce in Sweden, an epidemiological investigation was carried out. This revealed that the source of the infection was defatted thyroid powder, imported from Hungary, which was very heavily contaminated with *S. bareilly*, *S. muenchen* and faecal organisms (more than 3×10^7 cfu g^{-1} of powder). Further details of this interesting investigation were given by Kallings (1973). Coincidentally, another contaminated biological material, carmine, had been held responsible for outbreaks of salmonellosis in the USA (Lang *et al.* 1967, Komarmy *et al.* 1967, Eikhoff 1967). Capsules of this red dye, extracted from the cochineal insect and used to measure transit times of intestinal contents, were found to carry *S. cubana*. Eventually the potential danger was simply overcome by steaming batches of carmine for a short time before use. Two babies with fibrocystic disease of the pancreas were infected with *S. agona* after treatment with pancreatin, an enzyme-containing preparation of pig pancreas (Glencross 1972). No salmonellae were isolated from 0.5-g samples of the powder used but three serotypes, *S. agona*, *S. brandenburg* and *S. anatum*, were found when 10-g amounts were tested. Identical unopened bottles of the powder also yielded these organisms when similar amounts were examined. Later, *S. schwarzengrund* was isolated from two other children with fibrocystic disease and this uncommon serotype and *S. eimsbuettel* was found in batches of pancreatin used in a number of hospitals (Rowe and Hall 1975). Pig pancreas collected at abattoirs can be indetectably contaminated with salmonellae, and the gentle process used to obtain the enzyme-rich powder may not eradicate them completely, even if tests on 25-g amounts are negative. Glencross reported that the manufacturer had introduced a treatment with hypochlorite at an early stage in the production process with the aim of eradicating contaminating salmonellae. If bacterial multiplication is allowed to occur before administration of the powder, perhaps when it is added to a vehicle such as milk, an infective dose of bacteria may result from a very small number of contaminants.

2.2.1.3 The respiratory tract

Aqueous non-sterile preparations contaminated with Gram-negative bacteria have been found responsible for a number of respiratory infections in hospitals. Phillips (1966) traced an outbreak of respiratory tract infection to the lubrication of endotracheal tubes with a lignocaine jelly contaminated with *P. aeruginosa*. Aerosolization of fluids contaminated with Gram-negative bacteria into the stream of gas delivered by inhalation therapy equipment in hospitals is known to have constituted an important mechanism for causing bacterial pneumonia. The particles delivered are small enough in size to be deposited in terminal lung units where host defence mechanisms may be overcome more easily. In reviewing the topic, Pierce and Sanford (1973) points out that small-volume Venturi nebulizers, used with breathing machines to supply medication, had not been frequently incriminated in nosocomial epidemics but contaminated drug solutions used with the equipment had caused pulmonary infections with *Klebsiella* or *S. marcescens* on occasions (Mertz *et al.* 1967, Sanders *et al.* 1970).

An outbreak of *P. cepacia* infection was also attributed to anaesthetic atomizers (Schaffner *et al.* 1973). The organism originated from tap water used to dilute the topical tetracaine and cocaine anaesthetics used.

2.2.1.4 Topical administration

Once the skin becomes damaged, the risk of infection is greatly increased and organisms contaminating non-sterile products designed for use on damaged skin are known to have taken advantage of such opportunity. For instance *P. aeruginosa* caused minor, but clinically apparent, infections when a commercially available steroid preparation was used after dilution with an oily cream. Unfortunately the preservative, chlorocresol, became diluted to ineffective levels owing to its partition into the oily phase. This allowed the contaminating organisms to multiply to dangerous levels and to infect a number of patients over several months (Noble and Savin 1966). In an orthopaedic unit, *P. aeruginosa* gained access to wounds, probably via water expressed from wet plaster bandages. The water in turn picked up the bacteria by passage through contaminated cellulose wadding in the dressing (Sussman and Stevens 1960).

Another pseudomonad, *P. multivorans*, was isolated from infected operation wounds in nine hospitalized patients (Bassett *et al.* 1970). The bacteria were thought to have reached the wounds during skin preparation for minor surgery with a dilution of a chlorhexidine–cetrimide disinfectant. Experimental studies revealed that the disinfectant supported the growth of more than 10^6 pseudomonads ml^{-1}. It had obviously become contaminated from the hospital piped water during dilution.

Zinc-based barrier creams used to treat bedsores in patients in long-stay geriatric wards were found to harbour the same strains of *P. aeruginosa* and *S. aureus* which were also isolated from patients' sores (Baird *et al.* 1979). Other topical medicaments, particularly an emulsifying ointment used mainly as a substitute for bath soap by eczematous patients, were also involved in the cross-contamination of patients in a hospital for diseases of the skin (Baird *et al.* 1980). In these situations, it is often difficult, if not impossible, to decide the sequence of transfer between patient, ward environment and non-sterile preparation.

There is considerable evidence that antiseptic solutions used on skin have also been contaminated, but relatively few reports of resulting harm have appeared in the literature (Savin 1967). Although the potential hazard has long been appreciated, clinical infections have been reported in comparatively recent times involving intrinsic contamination of commercial iodophor antiseptic solutions with *P. aeruginosa* and *P. cepacia* (Berkelman *et al.* 1984, Parrot *et al.* 1982). Pseudomonads isolated as biofilms from pipework used in the manufacture of povidone-iodine antiseptics have been implicated in bacteraemias following the use of such contaminated solutions (Bond *et al.* 1983, Anderson *et al.* 1983).

2.2.1.5 Other routes

The intestine may be considered more at risk from enteric pathogens transmitted via the rectum than by the oral route because protection is not afforded by gastric acid. Meyers (1960) and Steinbach *et al.* (1960) simultaneously reported the observation of faeces in barium sulphate suspensions after performing barium enemas. Only the enema tip was discarded between patients and so harmful organisms could have been transmitted in the suspension. In fact, rectal transmission has occurred with both typhoid fever (Gilbert 1938) and bacillary dysentery (Hervey 1929).

Infection of the urinary tract is said to have been the most common type of nosocomial infection (Neu 1978, Ringertz and Ringertz 1982) as Gram-negative bacteria of faecal origin may cause infection following instrumentation or catheterization of the urinary tract. However, occasionally, the invading organisms may originate from a contaminated pharmaceutical product. In this way, pseudomonads growing in a quaternary phenoxy–polyethoxy–ethanol cleansing germicidal solution in a commercially packaged catheter tray caused a number of urinary tract infections following urethral catheterization (Hardy *et al.* 1970).

2.2.2 Cosmetics and toiletries

It is well documented that microbial contamination of cosmetics and toiletries occurs and, sometimes, potential pathogens have been found (Table 2.2). These products are used worldwide by the 10^{12} units annually and so, at first sight, it appears remarkable that the number of infections attributed to them is so few. A survey of the results of micro-biological examinations carried out by 29 manufacturers on returned complaint samples of product units alleged to have given rise to adverse effects was carried out by the Cosmetic, Toiletry and Perfumery Association, the UK trade association, in 1974. Pathogens were not found in any sample and none of the alleged harmful effects could be attributed to microbial contamination (Spooner 1977). Those reports that have reached the literature are mainly confined to individuals who are compromised in some way. A serious tetanus neonatorum outbreak arose in New Zealand from talcum powder contaminated with *Clostridium tetani* (Tremewan, 1946; Hills, 1946) and new-born babies in hospital have become colonized and infected with *P. aeruginosa* originating from a contaminated detergent cleansing solution and a handcream (Cooke *et al.* 1970, Baird 1977). A contaminated handcream has also been responsible for a serious outbreak of septicaemia in an intensive care unit (Morse *et al.* 1967). The few other published reports of infection have also been concerned almost exclusively with incidents involving toiletries used for hand cleansing and care in hospitals (Ayliffe *et al.* 1969, Jarvis *et al.* 1979).

On consideration, it is perhaps not surprising that so few infections have resulted from contaminated cosmetics. These products cleanse, beautify or modify the appearance of a person by external application and are intended to exert a local action rather than to act systemically. Normally, they are not swallowed although, exceptionally, small amounts of bath preparations, toothpastes or lipsticks may be accidentally ingested. Almost all cosmetics and toiletries therefore come into contact with healthy skin and this constitutes a good physical barrier against infection. The skin microflora also provides considerable protection against infection under normal conditions. It varies from a few hundred bacteria per square centimetre on the drier areas to many millions per square centimetre in moist situations such as the toe web (Noble and Somerville 1974). It may be calculated

Table 2.2 Some organisms isolated from unopened cosmetics and toiletries

Clostridium tetani	*Pseudomonas aeruginosa*
Clostridium perfringens	*Pseudomonas* spp.
Staphylococcus aureus	*Alcaligenes* spp.
Candida albicans	*Enterobacter* spp.
Escherichia coli	*Proteus* spp.
Klebsiella pneumoniae	*Acinetobacter* spp.
Moraxella spp.	

that the contribution to this bioburden by even a heavily contaminated product is extremely small as, in very round terms, the average amount of cosmetic applied to the skin is of the order of only $1\,\mathrm{mg\,cm}^{-2}$ (Spooner and Corbett 1985). Even if all the contaminating organisms were potentially pathogenic, it has been shown that they would probably become established only with considerable difficulty unless the balance of the normal flora is disturbed by other factors (Marples 1976).

It should be remembered, however, that the eye area is highly susceptible to infection, particularly if the eye is even slightly damaged. Workers in the USA demonstrated that samples of used eye cosmetics contained variable numbers of micro-organisms representative of the eye flora, notably *S. epidermidis* (Ahearn *et al.* 1974, Wilson *et al.* 1975). A corneal infection with *Fusarium* sp., a rare aetiological agent, had previously suggested an association between the use of contaminated cosmetics and infections of the eye (Wilson *et al.* 1971). Infections subsided when use of the incriminated cosmetics was discontinued and most were cases of blepharitis, a mild disease. However, a number of serious corneal ulcers associated with the use of contaminated mascaras has been reported (Wilson and Ahearn 1977, Reid and Wood 1979). The presence of *P. aeruginosa* in the corneal lesion and in the sample of mascara employed was established but the organism was not found in unopened samples of the same brands of mascaras. It was suggested that infection occurred after the corneal epithelium was scratched with a contaminated mascara applicator. The preservative systems in these products were clearly not capable of controlling the challenge from some of the constituents of the eye flora encountered.

2.3   The spread of hospital-acquired infection

Many of the incidents described so far have involved infections in one or more patients arising from an intrinsically contaminated product, one that constitutes an independent reservoir of infection. If a non-sterile product allows the survival or, particularly, the multiplication of harmful organisms, it may also act as an intermediate reservoir of infection, as shown in Table 2.3. The transmitted organism may have originated from the patient himself, the environment or the nursing staff or, in the case of communal medicaments, another patient. Bassett (1971) used 43 outbreaks of sepsis due to Gram-negative bacteria to exemplify such common-source incidents which had occurred over the previous two decades.

Table 2.3   Non-sterile products as intermediate reservoirs of infection

Preparation	Infecting organism
Disinfectant	Gram-negative bacteria
Antiseptic	*Pseudomonas aeruginosa, Pseudomonas cepacia*
Detergent	*Pseudomonas aeruginosa*
Emulsifying ointment	*Pseudomonas aeruginosa*
Peppermint water	*Pseudomonas aeruginosa*
Mouthwash	*Pseudomonas aeruginosa*, coliforms
Mascara	*Fusarium* sp., *Pseudomonas aeruginosa*
Enema	*Salmonella typhi, Shigella* sp.

Because of their extensive hospital use, disinfectant solutions constituted a hazard serious enough to elicit at least two warning articles in the 1950s (Anon. 1958a, 1958b). Despite this attention, further incidents continued to occur from time to time (Anon. 1976); a recent review of these incidents has been carried out by Bloomfield (1988). Solutions of quaternary ammonium compounds, particularly benzalkonium chloride, have been frequently incriminated and have been associated with serious infection such as *S. marcescens* meningitis (Sautter *et al.* 1984). The solutions are particularly at risk from metabolically versatile Gram-negative bacilli which may degrade the active ingredient in solution at 'in-use' levels. Aqueous solutions of chlorhexidine and chlorhexidine–cetrimide have also been frequently contaminated with *Pseudomonas* spp. and, for instance, have been involved in an outbreak of infection in an infants' nursery (Burdon and Whitby 1967). Diluted pine and chloroxylenol disinfectants have probably also acted as reservoirs of infection (Baird and Shooter 1976). In this case the preparations were known to have become contaminated with *P. aeruginosa* in the hospital pharmacy and had been inadvertently issued to the wards.

Hands of staff and patients are a very important means of transmission of hospital infection. After washing, they can quickly become recontaminated with organisms isolated from soap dishes (Jarvis *et al.* 1979). The use of handcream from a communal dispenser can also be dangerous. For instance lanolin handcream, provided by nursing personnel themselves in an intensive care unit, led to a serious outbreak of septicaemia due to *Klebsiella pneumoniae* (Morse *et al.* 1967).

Detergent solutions contaminated with Gram-negative bacilli have also facilitated the spread of hospital infection (Ayliffe *et al.* 1965, Victorin 1967, Cooke *et al.* 1970).

Communal oral preparations can become involved in the spread of infection. There is circumstantial evidence that peppermint water acted as a vector of *P. aeruginosa* in a hospital ward (Shooter *et al.* 1969) and an outbreak of septicaemia in an oncology ward was caused by use of a thymol mouthwash contaminated with *P. aeruginosa* (Stephenson *et al.* 1984). An extensive microbiological survey of patients and environment in a newly refurbished intensive care unit highlighted the importance of the oral route in the colonization of patients by Gram-negative bacteria. Mouthwashes and feeds were found to be heavily contaminated with epidemic strains of coliforms. Episodes of cross-infection occurred in the unit although very limited clinical infection associated with the investigated strains were seen (Millership *et al.* 1986).

2.4   Spoilage

A spoiled product is one that has been rendered unfit for its intended use. Contamination by harmful organisms obviously fits this definition but spoilage in the popular sense is also an important consequence of contamination. It involves deleterious changes which may render non-sterile products objectionable or even inactive. Spoilage of health and beauty products in particular is sometimes considered to be mainly an economic matter, as they have an aura in the market place of purity and cleanliness. Discovery by the customer of an inappropriately situated mould colony, for instance, is not likely to encourage further purchases of a particular brand. The commercial consequences of writing off spoiled batches of a product is sometimes also severe, if not disastrous. However, in addition to the aesthetic aspects, spoilage creates a potential danger to the patient or user. Changes may occur during storage of a product indetectably contaminated during manufacture. This may also happen in use as pharmaceuticals, cosmetics and toiletries are subjected to a high degree of insult by the purchaser. Organisms are readily introduced when half-emptied bottles are kept for long periods or by drinking from the bottle; other undesirable practices include dipping fingers into the cosmetic cream, spitting into the eye-liner or spilling the bath water into the shampoo. Mixed flora, introduced into the product by whatever means, are often extremely versatile in their metabolic activities and can adapt to a broad range of environmental conditions. All classes of natural organic compounds are susceptible to degradation by one organism or another, and synthetic materials may also be attacked. Pharmaceuticals and beauty products are highly processed materials and contain an enormous range of natural synthetic materials. Their susceptibility to microbial attack may result in extensive growth of contaminants and lead to devastating chemical and physical changes in the product. A description of the susceptibility to contamination of pharmaceutical ingredients, surfactants, polymers, humectants, fats, oils and sweetening, flavouring and colouring agents, has been published by Beveridge (1983). Sometimes, growth of species may be sequential, one organism paving the way for another. For instance, growth of an aerobic organism such as a pseudomonad may be followed by that of an anaerobe. This sequence is well known in engineering cutting oils and can cause the oils to become blackened owing to the eventual proliferation of sulphate-reducing bacteria (Guyne and Bennet 1959). We have observed similar changes in products such as indigestion mixtures which may be transmuted from white to a startling black by this mechanism.

2.4.1   Aesthetic manifestations

If ambient conditions encourage microbial growth, changes in the product may be an inevitable consequence of the metabolism of the multiplying organisms. These changes may be detectable by one or more of the senses, as previously described (Smart and Spooner 1972).

2.4.1.1   Visible effects

Homogeneous or colonial growth in, or on, a product probably constitutes the most striking and frequent manifestation. Contaminants may be seen as a sediment, turbidity or pellicle in liquid products. On more solid preparations, coloured colonies may form. The appearance of large bright yellow micrococcus colonies on a white cream is an alarming sight, particularly if an entire batch of a new product launch is affected. This may result from use of a natural ingredient, such as dried egg, which can carry large numbers of organisms of this type if improperly treated. Contamination by *Pseudomonas* spp., which are so metabolically versatile, is notable for the fluorescent soluble blue–green to brown pigments produced. Colour changes due to alterations in product components themselves may result as a direct consequence of metabolism or, indirectly, because of alteration of parameters such as pH or reduction–oxidation. Even distilled water, which can readily pick up organisms between the still and the point of use, may contain large numbers of bacteria within a few days. Subsequently, a wide variety of bacteria, moulds and yeasts can give rise to visible signs of growth. The presence of added organic material greatly increases the chance of growth, and we have seen deposits or turbidity due to algae, moulds, bacteria or yeasts in a range of poorly preserved pharmacopoeial solutions. Growth is not so readily detected in suspensions, except at the surface, because of inherent opacity but preparations of this type can thin, separate, decolourize or change colour as a result of microbial contamination. The same sort of change can occur in emulsions which may become visibly heterogeneous owing to hydrolysis of the oil phase or changes in pH of the aqueous phase. Mould growth is one of the most common visible manifestations of spoilage of creams. Containers are often the source of such trouble. Spores in dust particles, if not properly removed, may germinate and grow on the inside of the filled containers. Condensation in large air spaces of jars, owing to fluctuating storage temperatures, may also provide suitable conditions for the growth of spores present initially, perhaps on the inner surface of the closure. Occasionally, anhydrous products, such as ointments and oils, exhibit the same phenomenon if there is accidental ingress of traces of water or a humid atmosphere. Mould growth may also be visible on the surface of powders and tablets if they have been stored under damp conditions. Products such as shampoos, which contain a surfactant, are particularly susceptible to contamination by Gram-negative water-borne bacteria. Slimy sediments, pellicles and discolouration may occur. The loss of lathering properties may also be visible. The inclusion of surfactants which may be easily degraded biologically, particularly by *Pseudomonas* spp., in order to eliminate waste-water problems is not entirely advantageous. In contrast, frothing may be caused by contaminants if their metabolism produces enough gas from fermentable ingredients such as glycerol to exceed its solubility.

2.4.1.2 Olfactory effects

A variety of aroma-producing bacteria have long been identified (Omelianski 1923). Often their unpleasant odours are combined in spoiled products and are particularly disastrous in cosmetics and toiletries which depend so much on their own specific perfumes. Geosmin, a strongly earth-smelling neutral oil, is produced by some actinomycetes (Gerber and Lechevalier 1965) but the aromatic elements responsible for the typical smell of mould do not appear to have been identified. An alcoholic odour obviously indicates spoilage by yeast.

2.4.1.3 Taste

Reports that oral products taste 'peculiar' are sometimes the first indication that spoilage has occurred. However, an unpleasant flavour may not always be considered significant by the patient, as medicines are traditionally believed to be disagreeable. Although taste is a highly perceptive sense, it is also very subjective and cannot be relied on to detect spoilage. Nevertheless, over 100 different compounds are known to be involved in the production of flavour by micro-organisms (Margalith and Schwartz 1970) and some of these alcohols, aldehydes, ketones and other small molecules will inevitably occur in spoiled products and affect the taste.

2.4.1.4 Tactile effects

The texture of topical products is vital to acceptability. Contaminated creams may become lumpy and changes in the viscosity of contaminated liquids can be detected when applied to the skin. Butler (1968) reported that fungal contamination of a cosmetic powder gave rise to growth dense enough to cause serious modification of its mechanical properties.

2.4.1.5   Audible effects

The growth of organisms, even in heavily contaminated products, cannot actually be heard, of course, but their gas production may give rise to readily detectable sounds. Tubes of glycerol-containing formulations have been known to burst following fermentation of this substrate. Osmophilic yeasts, which have adapted to inimical conditions in tins of malt extract, have also announced their unsuspected presence explosively in this way!

2.4.2   Toxicity

There is a rather more sinister aspect of spoilage if it is not detected organoleptically. Consumers may be harmed by the production of toxins or metabolites or by the inactivation of biologically active constituents in a formulation. The occurrence in practice of such theoretical hazards is not well documented but, in some cases, they have been demonstrated in the laboratory.

2.4.2.1   Toxins

Many microbiological metabolites possess pharmacological activity and some have been intensively investigated because they also have antibiotic activity. Other molecules with pharmacological activity are produced by common spoilage organisms such as species of *Aspergillus* and *Pseudomonas* (Matthews and Wade 1977). Well-established toxins such as pyrogens, lipopolysaccharide components of the cell walls of Gram-negative bacteria, may be present even when viable organisms are no longer detectable. This hazard in parenteral products is well understood; in non-sterile products the significance of pyrogens is doubtful. It must certainly be taken into account when products are known to have been contaminated with Gram-negative organisms, even if there are no viable survivors. Highly dangerous metabolites, the mycotoxins, are produced by certain fungi (Wogan 1975). There is ample evidence that they have contaminated foodstuffs, constituting a health hazard to man and his livestock, particularly in developing countries (Jarvis 1976). There appears to be no evidence of a similar hazard from oral pharmaceuticals, and mycotoxins have not been found in fungally contaminated herbal drugs when sought (Hitokoto *et al.* 1978). The use of good pharmaceutical manufacturing practice should avoid the risk of mycotoxiosis by controlling fungal contamination. This is particularly relevant to raw materials such as starch as it has been shown experimentally that pharmaceutical starches can support the formation of mycotoxins when inoculated with spores of toxigenic strains of *Aspergillus* (Fernandez and Genis 1979).

Most microbial toxins are complex molecules but simpler catabolic products, including organic acids and amines, may be produced in relatively large amounts. Fortunately their presence in highly contaminated products is likely to be detected organoleptically by the consumer. Irritancy attributable to fragments of dead microbes in topical products is not established but the possibility that foreign protein evokes an allergic contact dermatitis reaction or primary irritation, particularly to the eye, should be borne in mind.

2.4.3   *Degradation of active constituents*

Beveridge (1975, 1983) summarized reports of laboratory demonstrations by himself and others of the inactivation of a range of potent drugs and antimicrobial agents by a wide variety of micro-organisms. Penicillins can be rapidly destroyed by β-lactamases and some other antibiotics, preservatives and disinfectants are metabolized. Alkaloids, analgesics, thalidomide, barbiturates and steroids have also been shown to be susceptible to inactivation in this way.

Although micro-organisms clearly have this propensity to degrade active ingredients, the phenomenon has not been investigated to any extent with formulated products which have been contaminated, either adventitiously or experimentally. Wills (1958) isolated a species of *Penicillium* from a sample of syrup of tolu which smelled of toluene. The organisms grew on benzoic or cinnamic acid as a sole carbon source and produced a toluene-like product from the latter acid. Atropine in eye-drops can be destroyed by *Corynebacterium* and *Pseudomonas* spp. (Kedzia *et al.* 1961) and the transformation of hydrocortisone in a dermatological cream to a therapeutically different compound by a contaminant, *Cladosporium herbarum*, has been reported by Cox and Sewell (1968).

There is little evidence that patients have actually been affected, but it is clear that microbial contaminants have the potential to cause considerable harm to the patient by this means if the organisms are not controlled.

2.5   **Conclusions**

Despite the obvious difficulties in establishing cause and effect, there is ample evidence that contaminated non-sterile pharmaceutical products have been responsible for the colonization of patients and hospital-acquired clinical illness in the past. Parker (1984) pointed out the difficulty of assessing the contribution of such infections to the death of patients in hospital because of the effect of the underlying disease. Certainly, hospital-acquired infection has prolonged the stay of patients in hospital for a matter of weeks. In the early 1960s the interest of hospital doctors transferred from antibiotic-resistant staphylococcal strains to Gram-negative aerobic bacilli as causes of infection. By the 1970s these organisms had colonized patients who were receiving antibiotics and so became endemic in hospital wards. It has been calculated that most hospital-acquired infections involve the urinary tract, surgical wounds or the respiratory tract (Harris *et al.* 1984). The majority of these cases are due to those Gram-negative bacilli which have only become known as opportunistic pathogens through their association with compromised hosts (Noble and White 1969, von Graevenitz 1977). Because of their relative insensitivity to antimicrobial agents, their metabolic versatility and the ability to multiply extensively in moist situations, it is not surprising that these bacteria have been responsible for the majority of incidents involving non-sterile products.

The situation is not quite the same with cosmetics and toiletries because they are primarily intended for, and used by, healthy people. Apart from a small number of ophthalmic infections originating from contaminated mascaras, infections have not been reported from this source although sometimes a significant degree of contamination has been observed. There is, nevertheless, a potential hazard, particularly with respect to eye products and spoilage, and this is appreciated.

Legislation often follows mishap and this is certainly so in the field of pharmaceuticals. In the USA, almost 200 deaths due to poisoning by ethylene glycol, incorporated into a sulphonamide elixir, gave considerable powers to the Food and Drug Administration while in the UK the thalidomide disaster lead to the Medicines Act. These incidents did not involve microbes but deaths from microbial contaminated infusion fluids led ultimately to the recommendations by the Rosenheim Committee on improvements in standards for the production of sterile products, and some non-sterile products, in the pharmaceutical industry and in hospitals (Rosenheim 1973). Nevertheless, there are still no international statutory microbiological standards for non-sterile products (Spooner 1985), perhaps partly because there have been no known mishaps comparable with those recorded for sterile products. Legislation covering the microbiological quality of cosmetics and toiletries specifically does not exist in most countries. In the USA this is largely due to recognition by the Cosmetic, Toiletry and Fragrance Association, the main trade association, of the potential hazard. This body adopted guidelines incorporating specific microbiological limits in 1973, and cognizance was taken by the Food and Drug Administration who felt it unnecessary to take specific statutory powers (Madden and Jackson 1981). Likewise, in Europe there are no legal standards at present although there is growing pressure for this to be changed in certain quarters.

Perusal of the literature suggests that the number of mishaps resulting from contaminated non-sterile products has diminished. This may be because the microbiological quality of industrial and hospital products has improved, as suggested in a number of publications (Ringertz and Ringertz 1982, Baird 1985) and also because there is a greater awareness of the hazards by the industry, the hospitals and the regulatory bodies. Nevertheless, as this review indicates, occasional mishaps continue to occur and there is no room for complacency. Improvements are needed in preservation and the 'first-generation' laboratory tests employed to measure its adequacy (Spooner and Croshaw 1981). Poorly formulated or unhygienically produced products may still put patients at risk, and communally used medicaments in hospitals should be regarded with suspicion. Those responsible for the production, distribution and administration of medicines must remain eternally vigilant. Meanwhile awareness of hospital-acquired viral diseases is increasing because of advances in our ability to culture these infectious agents and to perform diagnostic serology. Perhaps this suggests another problem just around the corner.

Acknowledgements

The author is considerably indebted to Margaret Hurst and John Spooner for help with the manuscript.

References

AHEARN, D. G., WILSON, L. A., REINHARDT, D. J., and JULIAN, A. J. (1974) Microbial contamination in eye cosmetics. Contamination during use. *Dev. Ind. Microbiol.* **15** 211–214.

ANDERSON, R. L., BERKELMAN, R. L., and HOLLAND, B. W. (1983) Microbiologic investigations with iodophor solutions. Proc. Int. Symp. Povidone, University of Kentucky Press, Lexington, KY, pp. 146–157.

ANON. (1958a) Failures of disinfectants to disinfect. *Lancet* **ii** 306.

ANON. (1958b) Bacteria in antiseptic solutions. *Br. Med. J.* **2** 436.

(1968) *Medicines Act*. HMSO, London.

(1971a) Microbial contamination in pharmaceuticals for oral and topical use: Society's working party report. *Pharm. J.* **207** 400–402.

(1971b) Microbial contamination of medicines administered to hospital patients: report by Public Health Laboratory Service Working Party. *Pharm. J.* **207** 96–99.

(1976) Benzalkonium chloride: failures as an antiseptic. *J. Am. Med. Assoc.* **236** 2433.

AYLIFFE, G. A. J., LOWBURY, E. J. L., HAMILTON, J. G., SMALL, J. M., ASHESHOV, E. A., and PARKER, M. T. (1965) Hospital infections with *Pseudomonas aeruginosa* in neurosurgery. *Lancet* ii 365–368.

AYLIFFE, G. A. J., BARRY, D. R., LOWBURY, E. J. L., ROPER-HALL, M. J., and WALKER, W. M. (1966) Postoperative infection with *Pseudomonas aeruginosa* in an eye hospital. *Lancet* i 1113–1117.

AYLIFFE, G. A. J., BARROWCLIFF, D. F., and LOWBURY, E. J. L. (1969) Contamination of disinfectants. *Br. Med. J.* **1** 505–511.

BAIRD, R. M. (1977) Microbial contamination of cosmetic products. *J. Soc. Cosmet. Chem.* **28** 17–20.

(1981) Drugs and cosmetics. In: Rose, A. H. (ed.), *Microbial biodeterioration*. Academic Press, London, pp. 387–426.

(1985) Microbial contamination of pharmaceutical products made in a hospital pharmacy: a nine year survey. *Pharm. J.* **231** 54–56.

BAIRD, R. M., and SHOOTER, R. A. (1976). *Pseudomonas aeruginosa* infections associated with use of contaminated medicaments. *Br. Med. J.* **2** 349–350.

BAIRD, R. M., FARWELL, J. A., STURGISS, M., AWAD, Z. A., and SHOOTER, R. A. (1979) Microbial contamination of topical medicaments used in the treatment and prevention of pressure sores. *J. Hyg.* **83** 445–450.

BAIRD, R. M., AWAD, Z. A., SHOOTER, R. A., and NOBLE, W. C. (1980) Contaminated medicaments in use in a hospital for diseases of the skin. *J. Hyg.* **84** 103–107.

BASSETT, D. C. J. (1971) Causes and prevention of sepsis due to Gram-negative bacteria: common sources of outbreaks. *Proc. R. Soc. Med.* **64** 980–986.

BASSETT, D. C. J., STOKES, K. J., and THOMAS, W. R. G. (1970) Wound infection due to *Pseudomonas multivorans*. A water-borne contaminant of disinfectant solutions. *Lancet* i 1188–1191.

BERKELMAN, R. L., ANDERSON, R. L., DAVIS, B. J., HIGHSMITH, A. K., PETERSEN, N. J., BONO, W. W., COOK, E. H., MACKEL, M. S., FAVERO, M. W., and MARTONE, W. J. (1984) Intrinsic bacterial contamination of a commercial iodophor solution. *App. Environ. Microbiol.* **47** 752–756.

BEVERIDGE, E. G. (1975) The microbial spoilage of pharmaceutical products. In: Lovelock, D. W., and Gilbert, R. J. (eds), *Microbial aspects of the deterioration of materials*. Academic Press, London, pp. 213–235.

(1983) Microbial spoilage and preservation of pharmaceutical products. In: Hugo, W. B., and Russell, A. D. (eds), *Pharmaceutical microbiology*, Blackwell Scientific, Oxford, pp. 334–353.

BLOOMFIELD, S. F. (1988) Biodeterioration and disinfectants. In: Houghton, P. R., Smith, R. N., and Egins, H. O. W. (eds), *Biodeterioration* Vol. 7. Elsevier Applied Science, Barking, pp. 135–145.

(1990) Microbial contamination: spoilage and hazard. In: Denyer, S. P., and Baird, R. M. (eds), *Guide to microbiological control in pharmaceuticals*. Ellis Horwood, Chichester. pp. 29–52.

BOND, W. W., FAVERO, M. S., PETERSON, N. J., and COOK, E. J. (1983) Observations on the intrinsic bacterial contamination of iodophor germicides. *Proc. Ann. Met. Amer. Soc. Microbiol.* **0** Q-133.

BRUCH, C. W. (1972) Possible modification of *USP* microbial limits and test. *Drug Cosmet. Ind.* **110** 32–37, 116–121.

BURDON, D. W., and WHITBY, J. L. (1967) Contamination of hospital disinfectants with *Pseudomonas* species. *Br. Med. J.* **2** 153–155.

BUTLER, N. J. (1968) The microbiological deterioration of cosmetics and pharmaceutical products. In: *Biodeterioration of materials*. Elsevier, Amsterdam, pp. 269–280.

COOKE, E. M., SHOOTER, R. A., O'FARRELL, S. M., and MARTIN, D. R. (1970) Faecal carriage of *Pseudomonas aeruginosa* by newborn babies. *Lancet* **ii** 1045–1046.

COX, P. H., and SEWELL, B. A. (1968) The metabolism of steroids by *Cladosporium herbarum. J. Soc. Cosmet. Chem.* **19** 461–467.

EIKHOFF, T. C. (1967) Nosocomial salmonellosis due to carmine. *Ann. Intern. Med.* **66** 813–814.

FERNANDEZ, G. S., and GENIS, M. Y. (1979) The formation of aflatoxins in different types of starches for pharmaceutical use. *Pharm. Acta Helv.* **54** 78–81.

GERBER, N. N., and LECHEVALIER, H. A. (1965) Geosmin, an earth-smelling substance isolated from *actinomycetes. Appl. Microbiol.* **13** 935–939.

GILBERT, J. (1938) Transmission of incidence of enteric disease by unsterile equipment used for administration of fluid by rectum. *J. Am. Med. Assoc.* **110** 1664–1666.

GLENCROSS, E. J. G. (1972) Pancreatin as a source of hospital-acquired Salmonellosis. *Br. Med. J.* **2** 376–378.

GREENWOOD, M. H., and HOOPER, W. L. (1982) Personal communication.
  (1983) Chocolate bars contaminated with *Salmonella napoli*: an infectivity study. *Br. Med. J.* **286** 1394.

GROVES, M. J. (1973) In: *Parenteral products*. William Heinemann Medical Books, London, pp. 7–10, 293–296.

GUYNE, C. J., and BENNET, E. O. (1959) Bacterial deterioration of emulsion oils. *Appl. Microbiol.* **7** 117.

HARDY, P. C., EDERER, G. M., and MATSEN, J. M. (1970) Contamination of commercially packaged urinary catheter kits with the pseudomonad EO-1. *New Engl. J. Med.* **282** 33–35.

HARRIS, A. A., LEVIN, S., and TRENHOLME, G. M. (1984) Selected aspects of nosocomial infections in the 1980s. *Am. J. Med.* **77** 3–10.

HERVEY, C. R. (1929) Series of typhoid fever cases infected by rectum. *Am. J. Publ. Health* **19** 166–171.

HILLS, S. (1946) The isolation of *Cl. tetani* from infected talc. *New Zealand Med. J.* **45** 419–423.

HITOKOTO, H., MOROZUMI, S., WAUKE, T., SAKAI, S., and KURATA, H. (1978) Fungal contamination and mycotoxin detection of powdered herbal drugs. *Appl. Environ. Microbiol.* **36** 252–256.

HOLMES, C. J., and ALLWOOD, M. C. (1979) The microbial contamination of intravenous infusions during clinical use. *J. Appl. Bacteriol.* **46** 247–267.

JARVIS, B. (1976) Mycotoxins in foods. In: Skinner, S. A., and Carr, J. G. (eds) *Microbiology in agriculture, fisheries and food*. Academic Press, London, pp. 251–267.

JARVIS, J. D., WYNNE, C. D., ENWRIGHT, L., and WILLIAMS, J. D. (1979) Handwashing and antiseptic-containing soaps in hospital. *J. Clin. Pathol.* **32** 732–737.

KALLINGS, L. O. (1973) Contamination of therapeutic agents. In: *Contamination in the manufacture of pharmaceutical products*. Secretariat of the European Free Trade Association, pp. 17–23.

KALLINGS. O., RINGERTZ, O., SILVERSTOLPE, L., and ERNERFELDT, F. (1966) Microbiological contamination of medical preparations. *Acta Pharm. Suec.* **3** 219–228.

KEDZIA, W., LEWON, J., and WISMIENSKI, T. (1961) The breakdown of atropine by bacteria. *J. Pharm. Pharmacol.* **13** 614–619.

KOMARMY, L. E., OXLEY, M., and BRECHER, G. (1967) Acquired salmonellosis traced to carmine dye capsules. *New Engl. J. Med.* **276** 850–852.

LANG, D. J., KUNZ, L. J., MARTIN, A. R., SCHROEDER, S. A., and THOMSON, L. A. (1967) Carmine as a source of nosocomial salmonellosis. *New Engl. J. Med.* **276** 829–832.

McCULLOUGH, N. B., and EISELE, C. W. (1951) Experimental human salmonellosis. *J. Infect. Dis.* **88** 278–289.

MADDEN, J. M., and JACKSON, G. J. (1981) Cosmetic preservations and microbes: viewpoint of the Food and Drug Administration. *Cosmet. Toilet.* **96** 75–77.

MALCOLM, S. A. (1976) The survival of bacteria in toiletries. In: Skinner, F. A., and Hugo, W. B. (eds), *Inhibition and inactivation of vegetative microbes*. Academic Press, London, pp. 305–315.

MARGALITH, P., and SCHWARTZ, Y. (1970) Flavour and micro-organisms. *Adv. Appl. Microbiol.* **12** 35–42.

MARPLES, R. R. (1976) Local infections – experimental aspects. *J. Soc. Cosmet. Chem.* **27** 449–457.

MATTHEWS, H. W., and WADE, B. W. (1977) Pharmacologically active compounds from microbial origins. *Adv. Appl. Microbiol.* **21** 269–288.

MERTZ, J. J., SCHARER, L., and McCLEMENT, J. H. (1967) A hospital outbreak of *Klebsiella pneumoniae* from inhalation therapy with contaminated aerosol solutions. *Am. Rev. Respir. Dis.* **95** 454–460.

MEYERS, P. H. (1960) Contamination of barium enema apparatus during use. *J. Am. Med. Assoc.* **174** 129–130.

MILLERSHIP, S. E., PATEL, N., and CHATTOPADHYAY, B. (1986) The colonization of patients in an intensive treatment unit with Gram-negative flora: the significance of the oral route. *J. Hosp. Infect.* **7** 226–235.

MORSE, L. J., WILLIAMS, H. I., GRENN, F. P., ELDRIDGE, E. F., and ROTTA, J. R. (1967) Septicaemia due to *Klebsiella pneumoniae* originating from a handcream dispenser. *New Engl. J. Med.* **277** 472–473.

NEU, H. C. (1978) Clinical role of *Pseudomonas aeruginosa*. In: Gilardi, G. L. (ed.), *Glucose nonfermenting Gram-negative bacteria in clinical microbiology*. CRC Press, West Palm Beach, FL, pp. 83–104.

NOBLE, W. C., and SAVIN, J. A. (1966) Steroid cream contaminated with *Pseudomonas aeruginosa*. *Lancet* **i** 347–349.

NOBLE, W. C., and SOMERVILLE, D. A. (1974) In: *Microbiology of human skin*. W. B. Saunders, London.

NOBLE, W. C., and WHITE, P. M. (1969) Pseudomonads and man. *Trans. St John's Hosp. Dermatol. Soc.* **55** 202–208.

NORTON, D. A., DAVIES, J. G., RICHARDSON, N. E., MEAKIN, B. J., and KEALL, A. (1974) The antimicrobial efficiencies of contact lens solutions. *J. Pharm. Pharmacol.* **26** 841–846.

OMELIANSKI, V. C. (1923) Aroma-producing micro-organisms. *J. Bacteriol.* **8** 393–415.

PARKER, M. T. (1972) The clinical significance of the presence of micro-organisms in pharmaceutical and cosmetic preparations. *J. Soc. Cosmet. Chem.* **23** 415–426.

(1984) Hospital acquired infections. In: Wilson, G., Miles, A., and Parker, M. T. (eds), *Topley and Wilson's Principles of bacteriology, virology and immunity*, Vol. 3. Edward Arnold, London, pp. 280–310.

PARROT, P. L., TERRY, P. M., WHITWORTH, E. N., FRAWLEY, L. W., COBLE, R. S., WACHSMITH, I. K., and McGOWAN, J. E. (1982) *Pseudomonas aeruginosa* associated with contaminated polaxamer-iodine solutions. *Lancet* **ii** 683–684.

PIERCE, A. K., and SANFORD, J. P. (1973) Bacterial contaminations of aerosols. *Arch. Intern. Med.* **131** 156–159.

PHILLIPS, I. (1966) Postoperative respiratory tract infection with *Pseudomonas aeruginosa* due to contaminated lignocaine jelly. *Lancet* **i** 903–906.

REID, F. R., and WOOD, T. O. (1979) Pseudomonas corneal ulcer. The causative role of contaminated eye cosmetics. *Arch. Ophthalmol.* **97** 1640–1641.

RICHARDSON, N. E., MEAKIN, B. J., and NORTON, D. A. (1979) Containers, preservatives and contact lens solutions. *Pharm. J.* **223** 462–464.

RINGERTZ, O., and RINGERTZ, S. H. (1982) The clinical significance of microbial contamination in pharmaceutical products. *Adv. Pharm. Sci.* **5** 201–225.

ROSENHEIM (1973) In: *Report on the prevention of microbial contamination of medicinal products.* HMSO, London.

ROWE, B., and HALL, M. L. (1975) *Salmonella* contamination of therapeutic panel preparations. *Br. Med. J.* **2** 51.

SANDERS, C. V., LUBY, J. B., JOHANSON, W. G., BARNETT, J. A., and SANFORD, J. P. (1970) *Serratia marcescens* infections from inhalation therapy medications: nosocomial outbreak. *Ann. Intern. Med.* **73** 15–21.

SAUTTER, R. L., MATTMAN, L. H., and LEGASPI, R. C. (1984) *Serratia marcescens* meningitis associated with a contaminated benzalkonium chloride solution. *Infect. Control* **5** 223–225.

SAVIN, J. A. (1967) The microbiology of topical preparations in pharmaceutical practice. 1. Clinical aspects. *Pharm. J.* **199** 285–288.

SCHAFFNER, W., REISIG, G., and VERRALL, R. A. (1973) Outbreak of *Ps. cepacia* infections due to contaminated anaesthetics. *Lancet* **i** 1050–1051.

SHOOTER, R. A., COOKE, E. M., GAYA, H., KUMAR, P., PATEL, N., PARKER, M. T., THOM, B. T., and FRANCE, D. R. (1969) Food and medicaments as possible sources of hospital strains of *Pseudomonas aeruginosa. Lancet* **i** 1227–1231.

SOMERVILLE, P. C., and SUMMERS, R. S. (1979) The clinical implications of contaminated pharmaceutical products. *Cent. Afr. J. Med.* **25** 195–198.

SMART, R., and SPOONER, D. F. (1972) Microbiological spoilage in pharmaceuticals and cosmetics. *J. Soc. Cosmet. Chem.* **23** 721–737.

SPOONER, D. F. (1977) Microbiological aspects of the EEC Cosmetics Directive. *Cosmet. Toilet.* **92** 42–51.

(1985) Microbiological criteria for non-sterile pharmaceuticals. *Manuf. Chem.* **56** No. 4, 41–45, No. 5, 71–75.

SPOONER, D. F., and CORBETT, R. J. (1985) The rational use of preservatives in cosmetics and toiletries. *Proceedings of Symposium on Formulating Better Cosmetics* 25–27 March 1985. Society of Cosmetic Chemists, London, pp. 1–13.

SPOONER, D. F., and CROSHAW, B. (1981) Challenge testing – the laboratory evaluation of the preservation of pharmaceutical preparations. *Pharm. Weekbl.* **116** 245–251.

STEINBACH, H. L., ROUSSEAU, R., MCCORMACK, K. R., and JAWETZ, E. (1960) Transmission of enteric pathogens by barium enema. *J. Am. Med. Assoc.* **174** 1207.

STEPHENSON, J. R., HEAD, S. R., RICHARDS, M. A., and TABAQCHALI, S. (1984) Outbreak of septicaemia due to contaminated mouthwash. *Br. Med. J.* **289** 1584.

SUSSMAN, M., and STEVENS, J. (1960). *Pseudomonas pyocyanea* wound infection: an outbreak in an orthopaedic unit. *Lancet* **ii** 734–736.

SYKES, G. (1971) The control of microbial contamination in pharmaceutical products for oral and topical use – raw materials. *J. Mond. Pharm.* **14** 8–20.

TEMPLETON, W. C., EIFERMAN, R. A., SNYDER, J. W., MELO, J. C., and RAFF, M. J. (1982) *Serratia* keratitis transmitted by contaminated eyedroppers. *Am. J. Ophthalmol.* **93** 723–726.

TREMEWAN, H. C. (1946) Tetanus neonatorum in New Zealand. *New Zealand Med. J.* **45** 312–313.

VICTORIN, L. (1967) An epidemic of otitis in newborns due to infection with *Pseudomonas aeruginosa. Acta Paediatr. Scand.* **56** 344–348.

VON GRAEVENITZ, A. (1977) The role of opportunistic bacteria in human disease. *Annu. Rev. Microbiol.* **31** 447–471.

WILLIAMS, R. E. O. (1967) Spread of airborne bacteria pathogenic for man. In: Gregory, P. H., and Monteith, J. L. (eds), *Airborne microbes.* Society of General Microbiology, Cambridge, pp. 268–285.

WILLS, B. A. (1958) Fungal growth in syrup of tolu. *J. Pharm. Pharmacol.* **120** 302–305.

WILSON, L. A., and AHEARN, D. G. (1977) *Pseudomonas*-induced corneal ulcers associated with contaminated eye mascaras. *Am. J. Ophthalmol.* **84** 112–119.

WILSON, L. A., KUEHNE, J. W., HALL, S. W., and AHEARN, D. G. (1971) Microbial contamination in ocular cosmetics. *Am. J. Ophthalmol.* **71** 1298–1302.

WILSON, L. A., JULIAN, A. J., and AHEARN, D. G. (1975) The survival and growth of microorganisms in mascara during use. *Am. J. Ophthalmol.* **79** 596–601.

WOGAN, G. N. (1975). Mycotoxins. *Annu. Rev. Pharmacol.* **15** 437–451.

PART TWO

Control in manufacture

3

Microbiological control of raw materials

M. RUSSELL

Centocor BV, Einsteinweg 100, 2333 CB, Leiden, The Netherlands

Summary

(1) The microbiological quality of raw materials can profoundly affect the quality of the finished products.

(2) To achieve the desired quality of the finished product, the raw materials must be monitored and controlled. This is performed by sampling and testing the raw materials to comply with appropriate specifications.

(3) Treatment of the raw material is sometimes necessary to achieve appropriate quality levels.

3.1 Introduction

Raw materials are here defined as those substances which are brought in either for further processing or to aid in such processing. Packaging materials are not included.

In any control system the following questions must be asked. Why is control attempted, what is being controlled, what is the aim of the control and what parameters are to be used in achieving this aim? Consequently, it must be decided which methods will best answer these questions and whether they are available. In the microbiological control of pharmaceutical raw materials there is one primary aim – to exclude any organisms which may subsequently result in deterioration of the product or may harm the patient. This aspect is discussed in detail in chapter 2.

Although instances of spoilage and infection are few relative to the total consumption of pharmaceutical, cosmetic and toiletry products, the fact that they have occurred is sufficient warning that the microbiological control over products must be exercised. A subsidiary aim is the control of microbial quality of the raw materials used in product manufacture. Contaminants in raw materials are important not only because they may contaminate the product directly but also because they may contaminate the manufacturing plant, possibly giving problems in the future. Experience has shown that, where free-living reservoirs of organisms become established within the manufacturing plant, they may be hard to eradicate and can result in continuous or intermittent contamination of the product. This may be difficult to detect.

In summary, the aim of microbiological control must be to give the manufacturer assurance that the raw materials used will not directly or indirectly represent a hazard to the product and ultimately to the consumer. It is very difficult to make a microbiologically clean product from microbiologically dirty raw materials.

In deciding suitable microbiological standards for raw materials, it is necessary to consider at the outset the use to which the raw materials are to be put. Not all products are intended or need to be sterile and the standard adopted must therefore reflect the type of raw material, the type of product, its method of manufacture and intended use (see section 3.2).

Mechanisms of control involve the sampling of a raw material and the testing of part of or all that sample for the presence of micro-organisms, with special reference to pathogenic species. Control also implies the use of antimicrobial treatments where appropriate or necessary. Some available methods will be discussed in section 3.5.

3.2   Raw materials requiring microbiological examination

The range of raw materials used in pharmaceutical and cosmetic products is extremely large and any attempt to sample and test every one would obviously prove too onerous. This has been tried, notably in the USA, but the resources necessary are disproportionate to the information obtained. Additionally, this blanket approach diverts attention away from the materials which are renowned to be potentially hazardous. Both experience and logic show that some raw materials are usually contaminated and that some never contain any detectable contamination (Russell, unpublished data). Obviously those in the latter category only need to be examined rarely or not at all, whereas those in the former category need regular examination. The criteria which may be used to identify materials to be examined are as follows.

(1)   Is the product of natural origin?

(2)   Is it synthetic?

(3)   Is it produced in a manner likely to reduce or increase any initial micro-organisms?

(4)   Are micro-organisms likely to multiply in it?

3.2.1   *Materials of natural origin*

Natural origin must be taken to mean any plant, animal or mineral material (e.g. gums, sugars, gelatine, hormones, talc or silica). Although methods of extraction may reduce the microbial population, there is always the possibility of contamination due to factors such as increased or changed initial population, change in procedures, or plant breakdowns.

The use of plant extracts in herbal remedies, cosmetics and toiletries is increasing in importance. These raw materials have a microbial population which is dependent upon several factors. These include the conditions of cultivation, the conditions of harvest, and the treatment and storage before processing: irrigation with farm slurry increases the coliform count; damp harvesting will cause the crop to be contaminated with organisms washed from the atmosphere by the rain, or splashed from the soil; the act of harvesting itself will cause mud to adhere more closely than dust. Drying, especially at higher temperatures, will reduce the microbial population but powdering or shredding will more uniformly distribute it.

Any increase in water activity (a_w) above the minimum required for growth will encourage population growth especially of moulds and fungi, as will insect or rodent infestation during storage. The latter may also bring about changes in the microbial population in local pockets which in turn may be difficult to detect by sampling.

W. McClean (personal communication) found that, of 20 herbal remedies surveyed, only six were considered satisfactory, i.e. had total counts of less than 10^3 cfu per gram, with no pathogens. The most common pathogens found were *Enterobacter* spp, which contaminated half of the samples tested. In another study the most common contaminants were moulds (*Aspergillus* and *Pencillium* spp) and spore forming aerobic bacteria (Payne 1990, 1993). However, coliforms and other Gram-negatives were also often found, as shown in Table 3.1. The levels of microbial contamination are known to vary from 10 cfu/g to 10^6 cfu/g (W. McClean personal communication, Hitokoto 1978). This variability may be due to the factors discussed above with regard to the harvesting and storage, or to the non-homogeneity of the sample.

Table 3.1 Microbial quality of tested excipients (adapted from McLean (1993, personal communication) and Payne (1990))

Substance	Microbial testing required by BP/EP (modified)	Comments (satisfactory to McLean's criteria)[a]
Gum acacia powder	Yes	Unsatisfactory – pseudomonads and coliforms
Alginic acid	Yes	Unsatisfactory – high count
Aluminium hydroxide	Yes	Satisfactory
Banana flavour		Satisfactory
Burdock extract powder		Satisfactory
Beta-carotene		Satisfactory
Calcium carbonate		Satisfactory
Calcium phosphate		Satisfactory
Caramel flavour		Satisfactory
Carmoisine soluble		Satisfactory
Celery seed		Unsatisfactory – high count and coliforms
Clivers extract		Unsatisfactory – high count
Cocoa powder		Unsatisfactory – high count
Dextrose		Satisfactory
Elderflower powder		Satisfactory
Gentian extract		Satisfactory
Ginger powder		Unsatisfactory – high count, Salmonella
Grapefruit flavour		Unsatisfactory – high count
Hops powder		Unsatisfactory – high count and coliforms
Hydrolysed gelatine	(Yes)	Unsatisfactory – high count
Hydrogenated glucose syrup		Satisfactory
Ispaghula husk		Unsatisfactory – high count, pseudomonads
Lactose	Yes	Satisfactory
Light kaolin		Satisfactory
Kelp		Unsatisfactory – Klebsiella spp.

(*continues*)

Table 3.1 (*continued*)

Substance	Microbial testing required by BP/EP (modified)	Comments (satisfactory to McLean's criteria)[a]
Locust bean gum		Unsatisfactory – high count
Lime flavour		Unsatisfactory – high count
Magnesium stearate		Borderline – high count
Magnesium trisilicate		Unsatisfactory – high count
Maize starch	Yes	Unsatisfactory – high count
Malt extract		Unsatisfactory – high count
Mannitol		Unsatisfactory – high count
Orange flavour		Satisfactory
Poke root powder		Unsatisfactory – high count and coliforms
Poplar bark		Satisfactory
Prickly ash powder		Unsatisfactory – high count and coliforms
Raspberry flavour		Satisfactory
Skullcap powder		Unsatisfactory – high count and coliforms
Senna pods		Unsatisfactory – high count
Sodium alginate	Yes	Unsatisfactory – high count
Sodium bicarbonate		Satisfactory
Sugar		Satisfactory
Sunset yellow		Satisfactory
Sterilized talc	(Yes)	Satisfactory
Uva ursi powder		Satisfactory
Yarrow powder		Unsatisfactory – high count and coliforms

[a] That is, total count $< 10^3$ cfu/g and no pathogens (McLean 1993, personal communication).

Some species have inherent anti-microbial properties which must be taken into account when testing.

If fungal growth on the material is suspected, then the presence and significance of fungal toxins must be evaluated.

Herbal remedies sold in the European Union require a product licence or marketing authorization (Anon. 1988). In order to assist in obtaining such an authorization, the UK Medicines Control Agency have issued Notes for Guidance on completing such an application. The notes require that vegetable drugs are tested for microbiological quality and that controls are exerted over their preparation. Additionally, the use of such raw materials in the factory must be controlled so that any microbial (or zoological) contamination is not allowed to spread, nor to contaminate other materials being processed. The *European Pharmacopoeia* has recently published microbial quality criteria for herbal remedies (Anon. 1995a).

In a similar way the *British Pharmocopoeia* (Anon. 1993a) and *European Pharmacopoeia* (Anon. 1983) lists raw materials of natural origin which must be tested for absence of certain specified micro-organisms and which must also comply with a total viable count limit in some cases. These requirements are summarized in Table 3.2.

Table 3.2   Raw materials with a microbiological specification in the *BP* (Anon. 1993a) and *European Pharmacopoeia* (Anon. 1983).

Raw material	Specification	Raw material	Specification
Acacia	1, 7	Lactose	3
Agar	7	Pancreatin	7, 10
Alginic acid	3, 4, 5, 8, 10	Pancreatic extract	1, 7, 10
Sodium alginate	2, 4, 5, 7	Pepsin	1, 7, 10
Aluminium	2, 7, 9	Polyacrylate 30% dispersion	1
hydroxide		Starch, maize, wheat,	2, 4, 6, 7
Bentonite	3	potato, rice	
Calcium gluconate	2	Starch, tapioca	7
Charcoal, activated	3	Sterculia	7
Cochineal	7, 10	Talc	3
Ferrous gluconate	3	Tragacanth	1, 7, 10
Gelatine	2, 4, 5, 7, 10	Trypsin	1, 4, 5, 7, 10
Heavy kaolin	3	(Shampoo)	3

Notes:
1. *EP* Some countries have a requirement for a total viable count of not greater than[a] 10^4/g or ml.
2. *EP* Some countries have a requirement for a total viable count of not greater than[a] 10^3/g or ml.
3. *EP* Some countries have a requirement for a total viable count of not greater than[a] 10^2/g or ml.
4. *EP* Some countries have a requirement for freedom from *E. coli* in 1 g or ml.
5. *EP* Some countries have a requirement for freedom from Salmonellae in 10 g or ml.
6. *EP* Some countries have a requirement for fungal count of not greater than[a] 10^2/g or ml.
7. *BP* 1 g or ml is free from *E. coli*.
8. *BP* 0.1 g or ml is free from *E. coli*.
9. *BP* 1 g is free from Enterobacteria.
10. *BP* 10 g is free from Salmonellae.

[a]   This is interpreted as the count not being greater than 5×10^x where $x = 2, 3$ or 4.

3.2.2   Synthetic materials

Synthetic compounds are rarely contaminated owing to the use of temperatures, organic solvents, extremes of pH, etc in the manufacturing processes. Although micro-organisms may contaminate all types of raw material, the potential for growth in synthetic materials is very low. In general, these incidental organisms are not detected by currently available microbiological techniques and are considered to be of little practical significance. Additionally contamination by pathogens is extremely unlikely. Exceptions to this rule can occur, for example, where one of the preparation stages involves washing with or crystallizing from water of poor microbiological quality. This may occur during the preparation of some lakes and dyes which may also be subsequently stored in water prior to use. From this, it can be seen that a knowledge of the method of preparation of the raw materials is essential in order to determine the extent of microbiological testing required for any material.

3.2.3 *Additional considerations*

It is often forgotten that micro-organisms need a minimum of 70% water in order to reproduce. Thus, multiplication in non-aqueous material is impossible. However, some materials, e.g. oils or sugar solution, may contain local pockets of free water where microbial proliferation may take place. In deciding whether to examine this type of product, clearly this factor must be considered.

Another criterion which may be considered is the reputation of the supplier and experience of the quality of material routinely supplied. Some suppliers realize the importance to the pharmaceutical and cosmetic industries of hygienic production of their product and will include a microbiological quality standard for their material which they will monitor. If their methods are satisfactory to the purchaser and their staff trustworthy, then there is little point in duplication of effort. In this case a certificate of analysis from the supplier will replace regular testing by the purchaser although spot or superficial checks may be helpful for quality assurance. The testing of incoming raw material by the purchaser is, in effect, end-product testing, which is acknowledged to give a lower level of confidence in the product, in comparison with an effective quality assurance programme.

The implementation of a reduced testing regime in house should only be considered when experience of testing every batch has been gained with one particular supplier (not raw material). In this situation, it may be appropriate either to test every batch of a particular material for total viable count or alternatively to perform the full range of tests, for example, on every fifth delivery.

If the raw material is consistently of bad quality, then the supplier should be approached to see whether improvements can be made. This should primarily involve improvements in good manufacturing practices which may include an antimicrobial treatment such as steam or ethylene oxide. Some raw materials may require the additional protection of a preservative mechanism, e.g. chemical preservative, or removal of free water. If these are not possible, then the choices for the manufacturer are few. He may change his supplier, include a antimicrobial treatment in his own process, change the raw material, reformulate his product or, on rare occasions, work to less stringent standards if the therapeutic benefits of the product are high enough. In this situation the classification of organisms into pathogens and non-pathogens must be done carefully, possibly with clinical advice.

Finally, it must be remembered that some material intended for parenteral products may be contaminated with pyrogens. Pyrogens in this context may be considered to be bacterial endotoxins or lipopolysaccharide (LPS) which is a component of the outer membrane of the cell envelope of viable and non-viable Gram-negative bacteria. Other pyrogens, such as those formed by algae, fungi and Gram-positive organisms are not a problem in practice. If the raw materials need to be free of pyrogens, then the Limulus Amoebocyte Lysate test should be considered, as it is cheap and does not require living animals for its performance. Details of the test are included in the European Pharmacopoeia (Anon. 1983).

3.3 Sampling

This is a most important area. The purpose of taking a sample is to judge the quality of the whole batch on the results of the tests performed upon the sample. This presupposes that microbial contamination is homogeneously dispersed throughout the material being sampled. This may or may not be true. In general, aqueous solutions and suspensions and their microbial flora are assumed to be homogeneous. Non-aqueous materials may not be assumed to be equally contaminated, as, for example, where oils become contaminated with small quantities of water. Therefore, although sampling theory normally demands a random sample, this may not be appropriate for the microbiological sample. If the process of manufacture is known, then samples may be taken from the known vulnerable points. This approach is known as hazard analysis of critical control points. If not, then random sampling has to be used but the interpretation of results, including the use of resampling is more difficult. Hazard analysis of critical control points, which is a system whereby the vulnerable points in a process are monitored to ensure the system is in control, is discussed in more detail in chapter 4.

The following discussion is based on that given in the books from the International Commission on Microbiological Specifications for Foods (ICMSF) (Silliker 1980a, 1980b). The questions to be asked about sampling are always how large a sample is necessary and how many samples should be taken: the answers to these questions depend on how sure the questioner needs to be about the accuracy of the results. The sampling rates required can be obtained from published tables (Anon. 1972). Examination of these, however, will show that the number of samples to be taken rises dramatically with the assurance required. Two methods have been adopted in practice. Although they are not mutually exclusive, they tend to be used as such. In the first method a sample is taken from each of a number of containers in a delivery. This number is usually the square root of the number of containers delivered, plus two. It appears that there is no statistical reason for this number (J. Altman, personal communication). These samples are bulked and mixed to give a composite sample. A portion of this sample is analysed and the delivery accepted or rejected on the results of the analysis, with resampling if necessary. The results of many such analyses can be examined for trends.

The second method, attribute sampling, is recommended for use in the cosmetic and food industries. As before, a number of samples are taken but are analysed separately. This gives a more complete picture of the status of the material and its acceptability is judged on all the results based on a prearranged plan. This method is much more expensive in terms of resources but, as referred to in the introduction, if the resources are targeted to the vulnerable materials, then it is more cost-effective.

With a two-class attribute sampling plan, the number n of samples to be taken and the maximum allowable number c of positive results are defined. Thus, with $n = 10$ and $c = 2$, 10 samples are taken and analysed and not more than two of them may be positive for the criterion laid down (e.g. counts above 500 colony-forming units per gram (cfu g^{-1})). Thus the two classes are above and below 500 cfu g^{-1}. The two-class attribute plan is adopted in the guidelines given in *The bacteriological examination of drinking water supplies* (Anon. 1982). *Escherichia coli* is not allowed in any sample and only two samples of any five may contain Enterobacteriaceae.

With the three-class system an intermediate level is introduced. Thus, in the example above, it could be that a count below 500 cfu g^{-1} is acceptable, and a count above 1,000 cfu g^{-1} is unacceptable, the level in between being marginally acceptable.

In general the first system is used where no positives are to be tolerated, e.g. freedom from Salmonellae in 25 g ($n = 5 \times 5$ g, $c = 0$) whereas the second would be used where counts are made and different levels of contamination are considered as appropriate. With the three-class attribute sampling system, trends are more easily monitored because only the intermediate value need be recorded and this is a concentration or a count, whereas in the other classes results are recorded as acceptable or not acceptable, thus giving less information. Sampling may be reduced after examination of a number of batches indicates satisfactory results. Implementation of reduced sampling may involve a reduction in the number of samples per lot or the number of lots tested or even the types or numbers of tests carried out on each sample. Reducing the number of samples will reduce the stringency or discrimination of the test although, if the lot size is large in comparison with sample size, this will not be of practical importance. In this case, every fourth or fifth delivery may be tested until a problem arises and then the sampling will revert to every delivery until confidence is rebuilt. This system obviously requires some organization – does the laboratory or the sampling organization keep the records?

A different approach to reduced testing may be adopted in some circumstances, e.g. only testing for specific organisms if the aerobic plate count exceeds a preset number. This presupposes that some specific organisms are allowable; see section 3.6.

Although each batch of material is released or rejected on its own merits, much more information, and therefore assurance of quality, can be gained if the results from many batches are combined. This may be done historically or concurrently. The latter is possibly easier to organize and therefore the more useful. The cumulative sum (cusum) chart is well recognized as a method of showing trends in a repeatedly monitored system, such as water monitoring (Anon. 1980a). Observation of the raw data rarely shows easily discernible trends, whereas cusum charts clearly display any trends. Thus, warning that the system is going out of control may be obtained before control is lost. Examples of this are given in Figs 3.1 and 3.2. Fig. 3.1 shows consecutive counts of deionized water plotted against day number. Fig. 3.2 shows the same data plotted as a cusum using a previously selected target of 24. The significance of the results is easier to interpret in Fig. 3.2 which demonstrates an event on day 15 which influences the microbial quality of the water. This event is maintained over at least the following 10 days. Calculation of cusums is easy but time-consuming. The target figure is selected by reference to the data and the quality standard. The cusum is calculated as follows:

cusum = (target – observation) + last cusum

This is plotted against the observation number (day number in Fig. 3.2). Thus the cusum varies according to the difference from the target and if the sign of the difference (+ or –) between the target and observation stays the same, then a trend will be apparent. Upward trends indicate a loss of control and vice versa. For a fuller discussion, refer to Russell *et al.* (1984).

Sampling for microbiology is rarely carried out by microbiologists. Therefore the people who actually take the samples must be trained in aseptic techniques so that they understand the ease with which samples may be contaminated and therefore invalidated. The necessity for resampling requires careful consideration. A resample must always be taken when there is doubt as to the validity of the sampling or test. Also the result of a high count may be worth re-examination owing to inaccuracies in the counting method, possible changing microbial populations, and non-homogeneity of the micro-organisms in the material. Of course, if a banned organism is found, then the material must be rejected. It is very important to agree the specification and methods with the supplier before testing.

3.4   Water

Water is the most widely used raw material. It may be used as an ingredient, cleansing

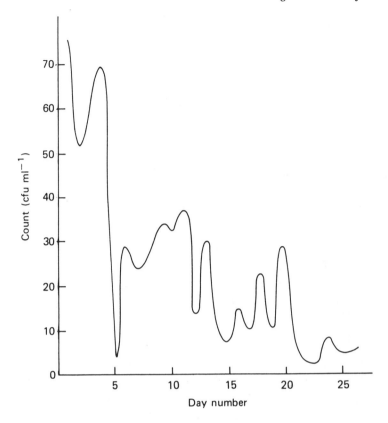

Figure 3.1 Bacterial counts obtained from deionized water plotted against day number. Trends and events are not easily observed

agent or temporary suspending agent. The methods of water treatment, its storage and distribution all have an influence on its microbial quality and all these factors must be taken into account when assessing its quality. Micro-organisms, particularly Gram-negative rods, can grow rapidly in water to 10^5–10^6 cfu ml^{-1} under all ambient conditions, clearly showing water to be a high-risk raw material. This level of contamination cannot be seen with the naked eye.

The types of water available for use are potable water (as defined by EEC Directive (Anon. 1980b)), distilled water, deionized water, demineralized water and water produced by reverse osmosis. Potable water may be used for all toiletries and cosmetics and some pharmaceuticals although it has the disadvantage that its quality may vary both from time to time and from place to place. Additionally, in the UK, water authorities will not allow water to be used directly from the mains but insist on break tanks. The use of break tanks may represent a significant source of microbial contamination because of the extended storage of the water.

Distilled water is very-high-quality water, if produced by a still designed to prevent the entrainment of water droplets, but its microbiological quality deteriorates quickly unless stored correctly. These storage conditions are generally accepted to be above 60°C or under aseptic conditions. Owing to its high cost, distilled water is usually used only for parenteral manufacture either as an ingredient or as a pyrogen-free rinsing agent for product contact surfaces.

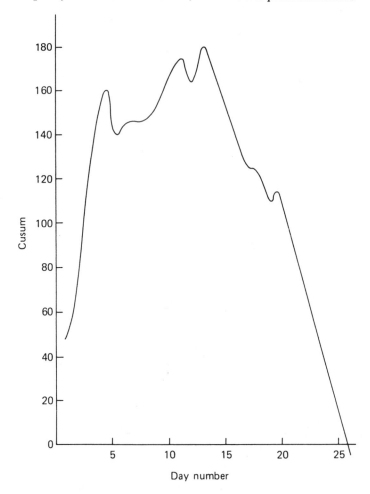

Figure 3.2   Cusum chart of the data presented in Fig. 3.1. The target value used is 24 and the trends and events in the data are easily discernible

Reverse-osmosis water can be of similar quality to distilled water but care must be taken to disinfect the membrane at regular intervals. This interval will be determined by the results of regular sampling but will probably be of the order of once per month, depending on use.

Deionized water is used extensively in the manufacture of tablets, syrups, suspensions, creams, lotions and for washing of all the manufacturing equipment. Problems may arise owing to either the deionizing bed or the distribution system. The deionizing bed may be controlled by frequent regeneration but this will not necessarily eliminate the contamination.

Deionization beds are prone to contamination because they must be protected from the chlorine which acts as a bacteriostat in potable water. A new technology rapidly gaining acceptance is continuous electrode-ionization (Gallantree 1994) where a direct electric field is used in combination with ion exchange membranes and resins to remove ions from water. This system also appears to reduce the numbers of micro-organisms, possibly since they act as charged particles.

Potable water must be tested to ensure compliance with the EEC Directive (Anon. 1980b) which means testing for coliforms, *E. coli* and total viable count.

The total bacterial count action limits expected by the US Food and Drug Administration are 50 cfu/ml (alert) and 100 cfu/ml (action) for purified water and 10 cfu/100 ml for water for injection (Anon. 1993b).

The Cosmetics, Toiletries and Perfumery Association recommends a limit of not more than 100/ml at the point of use (Anon. 1990).

Sampling and testing of water must be carried out appropriately for the use of the water, and the way in which it is prepared and stored. Deionized water should be sampled weekly or between regeneration cycles of the deionizing bed, whichever is the shortest. This assumes that no unacceptable bacterial build-up occurs during the preregeneration period. This will be established by collecting background data during commissioning and validating the system. Distilled water used for non-parenteral manufacture also requires frequent monitoring. With no regeneration to hold down microbial numbers, regular monitoring is essential to ensure that control is maintained.

The responsibility of the water authority for the quality of mains water ends at the boundary of the receiver's property. Valves and break tanks can be a major source of microbiological contamination. If mains water is used for any production purpose, then it must be regularly monitored, at least weekly.

For all types of water, samples should be taken from each outlet, as they may be differently contaminated, and also at each point of treatment (e.g. last deionizing bed, reverse-osmosis membrane, still, and break tank) so that the origin of any problem can be determined.

3.5 Treatment methods

Antimicrobial treatment of raw materials is difficult for the user company. Water is less problematical and is discussed separately. The options available for raw materials other than water are very limited and consist generally of heat, irradiation and ethylene oxide treatment. Other possibilities such as filtration and disinfection are generally not applicable owing to the nature of the material. Time may be useful. Depending upon the material and the microbial flora present, bacterial numbers often decline over a period of weeks. This is particularly true for non-aqueous raw materials but should be tested in each case if it is used as a method of control. It is important to note that treatment is usually performed to reduce bacterial numbers or to eliminate or reduce specific species of micro-organisms rather than sterilization. Most human pathogens are killed by heat at 60°C. Although bacterial spores and moulds are more resistant to heat, they are considered mainly for their spoilage potential.

The nature of the material is the determining factor in deciding which treatment is appropriate. Moist heat is not appropriate to treat starch or gelatine capsules. Dry heat is appropriate for inorganic materials such as talc and silica. Ethylene oxide is strongly reactive and is absorbed by some materials. Irradiation may be used but the costs (transport, treatment and inventory time) have to be carefully considered, as well as the possible deleterious effects on the material.

There is little information in the literature on the pasteurizing effect of dry heat on solid materials. Investigations have shown that 2 h at 80°C reduces Aspergillus spores by two logarithmic cycles (M. Russell, unpublished data). This treatment did not apparently denature the starch carrier but did bring the count within specification. Heat penetration both within the powder and within each particle has to be taken into account in any treatment. The same argument could apply to oils, waxes and gums; this dry heat treatment may be considered if the material will withstand the time–temperature regime necessary. This should be ascertained on an individual basis.

Many options are available for improvement of water quality. These are heat, ultraviolet irradiation, ozonolysis, filtration, frequent deionizer bed regeneration, recirculation and disinfection. Production of good-quality water also depends on correct plant design and operation. The use of any particular treatment method depends on what is causing the microbial deterioration, the source of the problem, the quality of water required and the volume to be treated, the type of distribution system, and the capital expenditure available.

The best-quality water is probably distilled water stored above 60°C and recirculated via a membrane filter through stainless steel pipes at high speed (greater than 2 m s^{-1}). This system would be drained, dried and steamed at weekly intervals. However, the cost of producing water of this quality is prohibitive for most areas.

Very few organisms will grow above 60°C and these are generally not detectable by normal methods (see below). Thus, heat at this temperature is a good water treatment, if the distribution system will allow it and the energy costs are acceptable.

Ultraviolet irradiation in a recirculating system works well and is relatively cheap to run. Caveats are that the penetration of ultraviolet light into water is small (of the order of 2 cm) which will reduce the flow through the cell and therefore impose a bottleneck on the system; any dead bacteria present in the system will further hinder the penetration of the ultraviolet light; the intensity of the lamp must be regularly monitored according to the manufacturer's instructions (see chapter 6, section 6.5.2). However, recent improvements in the design, where the water flows over the lamp and the use of multiple lamp systems, have increased efficiency considerably, especially if combined with filtration to remove dead bacteria.

Ozonolysis, as with ultraviolet light and heat, does not leave chemical substances in the water to be removed. However, the capital investment is high and the safety requirements are complex. Since improvements in design are making this treatment more accessible and cheaper, it is probably the system of the future.

Filtration works well but is relatively expensive for high throughputs because the filters may need regular changing to prevent blockages. Also energy costs may be high because of the necessity to force the water through the sterilizing grade filters.

Improvements to water systems have been made by combining several treatments, thus reducing the disadvantages of each treatment.

There is evidence that frequent regeneration of deionized beds reduces microbial numbers; at the design stage it may be worth installing two half-size units so that regular regeneration can be performed.

Recirculation at speeds greater than 2 m s^{-1} prevents bacterial multiplication in deionizers possibly by a scouring action on the walls of the pipes (Elgar Ltd, personal communication).

The design of a system is influential on the size of microbial populations and the ability of the user to remove them. Dead-legs, long pipework runs to taps, undrainable pipes and U-beds, heat-labile materials and organic packing all create microbiological problems once installed. Once contaminated, resin beds are very difficult to treat and the manufacturer must be consulted. Contamination of the bed may be avoided by short regeneration times and good-quality input water, e.g. the use of filtration or ultraviolet treatment prior to the deionizer.

The whole distribution system may be disinfected either chemically or by heat, if the pipework and fittings will withstand it. It must be ensured that all parts of the system are exposed to the disinfecting agent for a sufficiently long time. This may be impossible in older installations where dead-legs, long runs to taps and organic packing in fitments are often found. When treating the system, it should be remembered that bacteria are usually protected by extracellular slime from the effects of disinfectants. The penetration of this slime usually requires aggressive chemicals and long contact times. The slime may be sufficiently protective to allow resistance to the disinfectant to develop which may subsequently allow growth of the organism in the product, particularly in cosmetics. Chemicals in use are hydrogen peroxide, (6–10% for minimum of 6 h), chlorine (100 ppm for a minimum of 1 h), formaldehyde (2–4% for minimum of 1 h) and ozone (0.5–5 ppm for 0.5–1 h). The concentrations given must be maintained throughout the whole system. Chemicals must, of course, be fully flushed from the whole system, with appropriate-quality water. Adequate flushing is not easy in a poorly designed system.

If the system can be drained and dried, then the microbial population can be reduced, remembering that most water-borne bacteria are very susceptible to drying. The practicalities of this would have to be investigated in each case. It must be stressed that the whole system must be treated, since even small amounts of residual contamination will ultimately lead to the re-establishment of free-living reservoirs within the system.

3.6   Methods of testing

The methods used in testing must reflect the aims of the testing. If the numbers of organisms are important, then counting methods must be used. If the type of organism is important, then a method selective for that organism must be adopted. Usually both are used although in a reduced testing scheme only one may be necessary, e.g. if the count is persistently less than 100 g^{-1}, it seems unnecessary to test for specific organisms. If only one type of organism is a problem, e.g. *Pseudomonas* spp., then it may be necessary to look for only this, ignoring all other organisms and their numbers. The methods adopted must also reflect the nature of the material under test and, possibly, any pretreatment, such as chlorination of the water supply. Liquids may be filtered or dispersed in a nutrient medium and solids may be dissolved or dispersed and tested.

For some materials, e.g. waxes and other oily solids, it is difficult to devise meaningful tests and in these cases the effect of any microbial contamination on the final dosage form has to be examined. This is examined in more detail below.

3.6.1   Counting methods

Three types of method are available here: plate counts, filtered counts and use of multiple tubes (most probable number techniques). These all presuppose the feasibility of preparing a solution or dispersion of the test material. The effect of the material (e.g. osmotic effect of a sugar) on the counting method must be considered. Paradoxically, when counting micro-organisms, the number present must be known beforehand in order that a method of appropriate sensitivity may be chosen.

For plate counting a known aliquot of a solution, or suspension, of the material under test is added to molten agar and poured into Petri dishes, or added to the surface of a prepoured plate. At least duplicate plate counts must be prepared. The plates are incubated at a temperature and time appropriate for the type of micro-organism to be isolated, e.g. psychrophilic spoilage organisms must be incubated below 16°C. The resultant microbial colonies are counted.

Plate-counting methods are quick and easy to use but may be subject to counting errors especially if the wrong dilution of original material is chosen. Counting methods as a whole are inaccurate and 100% errors are common. The pour plate method does not count less than 6 organisms ml^{-1} of the original suspension, since (for statistical reasons) 30 is the minimum number of colonies per plate which may be counted and it is inconvenient to add more than 5 ml per pour plate without reducing the gel strength of the agar. Conversely, the upper limit is 3000 cfu ml^{-1} using a 0.1 ml inoculum, without dilution, since 300 colonies per plate is the maximum which may be counted. For the surface plate method the lower limit is 15 cfu ml^{-1} of original suspension, since it is not practical to add more than 2 ml to the surface of the plate. The Miles–Misra (Miles *et al.* 1938) technique is a useful variant, when drops of a known size (usually 25 microlitre) are placed on the surface of a predried plate. Fifteen drops may be placed on a plate and thus several dilutions can be counted on one plate. The plates are incubated for a shorter time than normal as only very small colonies are counted before they merge together. The colonies may be counted under a low-power microscope. Five duplicate drops are usually counted per dilution. Highly coloured and viscous materials can give problems with this method (for both counting and mixing reasons) as can colonies growing together when mobile or spreading forms are counted. These methods count viable particles or colony-forming units. It must be remembered that moulds may grow in a solution but do not form any colony-forming units.

Selective media and antimicrobial antagonists must be used with care, since both can adversely affect the number of colonies formed. Antimicrobial agents present in a sample must be inactivated to provide a true estimate of microbial numbers. This may be done by removal (see below), dilution or chemical inactivation. The inactivating agent must be specified for the preservative present (Russell *et al.* 1979). The time and temperature of incubation are also important and may have a selective effect on colony formation, particularly in a mixed population. Many water-borne organisms will not grow above 30°C or even above 25°C. *The bacteriological examination of drinking water supplies* (Anon. 1982) suggests 25°C for a minimum of 4 days. Knowledge of this time–temperature relationship is important in the interpretation of the results obtained. For example, many species of bacteria will grow on media selective for Enterobacteriaceae, if the medium is incubated for longer than 24 h.

Filtered counts are made by filtering a known volume of liquid through a membrane filter and then placing the filter on an agar plate (or nutrient pad) and incubating. Nutrients diffuse through the membrane, allowing any colonies to grow and be counted. The practical upper limit for this method is 100 organisms per aliquot. Any more than this and the colonies grow together and are uncountable. It is impracticable to use smaller amounts than about 5 ml; otherwise the colonies do not evenly cover the surface although, in this situation, dilutions can be made.

This technique allows highly coloured material and also antimicrobial agents to be washed away but it suffers from the significant drawback that only liquids and soluble solids may be tested. Some fats may be solubilized in appropriate oils (arachis, mineral and coconut) for filtering. Viscous liquids, such as oils, may be impracticable to test because of the filtration time. However, the membrane may be placed on either non-selective or selective media or, indeed, placed into a nutrient broth for enrichment. The major advantage of this method is that large volumes of liquids may be processed.

The most probable number technique appears to be little used in the pharmaceutical industry. In this method, known amounts of product and/or diluted product are added to a series of tubes, usually three or ten. The tubes contain a known amount of appropriate broth (usually 9 or 10 ml). The tubes are incubated and the number of tubes showing growth counted (Cruikshank 1965). By reference to published tables, the most probable number of organisms in the original sample can be calculated easily (Cruikshank 1965, Anon. 1982). This method is very useful and sensitive but it suffers from two major disadvantages; it is reliant on aseptic techniques and it is cumbersome. Coloured or turbid solutions are also a disadvantage, requiring either subculture or microscopy to demonstrate the presence of growth. This method is very sensitive for detecting small numbers of particular types of bacteria, e.g. coliforms, and the technique is used by the dairy and water industries for this purpose.

The dip slide method has been used to estimate the numbers of bacteria in a solution. In this technique, a slide coated with a nutrient agar (usually commercially obtained) is dipped into the solution to be tested and then incubated. This will give an approximate number of bacteria per millilitre of original solution. However, the minimum level of detection is 1000 cfu ml^{-1} and levels greater than this are not often detected in the samples from the pharmaceutical or cosmetic industries.

The Miles–Misra drop count (Miles *et al.* 1938) can be usefully employed as a screening test when relatively high viable counts are expected.

3.6.2 Methods for specific microbial types

The methods to be used to detect specific types depend entirely on the types of organism looked for. Therefore it has to be decided which organisms are of interest. The two classes to consider are pathogenic organisms and spoilage organisms. The first class includes those organisms which may produce topical infections when applied to the skin and mucous membranes, and those which cause intestinal infections when given orally. Organisms of interest for topical preparations include *Staphylococcus* spp., *Pseudomonas* spp. and *Candida albicans*. Organisms of interest for oral products are those which cause food poisoning. The range of food-poisoning organisms is increasing year by year but Salmonellae and Clostridia are probably still the most important. However, most companies would also check for the presence of *E. coli* as a classic indicator of human faecal pollution, and Enterobacteriaceae generally, as indicators of poor hygienic practice. *Bacillus cereus* is of increasing concern since its implication as a food-poisoning organism. *S. aureus* may also be of interest, both as a potential pathogen in this context and as an indicator of the quality of the hygienic manufacture of the material.

Spoilage organisms are very difficult to predict. Where a particular spoilage event has occurred, the specific organism may be looked for but, in the absence of such experience, counts of yeast, mould and Gram-negative water-borne organisms are usually taken as predictors of possible spoilage problems.

Practical details of methods for the detection, isolation and identification of such organisms can be found, for example, in the pharmacopoeias (Anon. 1983, 1993, 1995b), the ICMSF books (Silliker 1980a, 1980b), specific books on identification (Cowan 1993), specific industry guidelines (Anon. 1989, 1990) or from the manufacturers of microbiological culture media. It should be remembered that, in most cases, organisms in raw materials will not be growing. They may also be damaged or stressed by the production process. Therefore the direct inoculation onto selective media is not recommended. Selective media for a specific micro-organism are designed to be inhibitory to growth of other species but may also have some inhibitory effect on the selected species. Therefore a pre-enrichment stage in a general-purpose non-inhibiting environment is necessary. Subcultures are made after incubation in appropriate selective media. Selective media are not usually specific for a single species of organisms, allowing a class of organisms to grow. Further tests are usually required to confirm the identity of the suspected organism before rejection of the material is considered. These tests involve the evaluation of the biochemical reactions of the micro-organism under test. For some classes of organisms, this may be done with commercially available kits. The interpretation of the results of such tests is not easy and positive controls are always necessary for the inexperienced.

3.7 Conclusion

As discussed in the introduction to this chapter, raw materials do not need to be sterile but some form of microbiological control is necessary to ensure that neither the consumer nor the product is exposed to an unacceptable level of risk. It should be remembered that people take in large numbers of organisms in their food with no untoward effect and indeed in some cases with a beneficial effect. Also the rate of incorporation of a raw material in a product can have a significant diluting effect which could give an acceptable final product. Many processes include antimicrobial treatments, e.g. heat above 60°C or solvent extraction.

Therefore the level of control necessary must be judged in the context of the individual circumstances of the raw material and the product being manufactured. 'Acceptable risk' must be also defined in terms of the raw materials, the production process, the final product and its therapeutic use.

Acknowledgements

I would like to thank Mrs M. Evans and Mrs P. Fisher for their considerate help in the preparation of this manuscript and I. Gallentree and D. Hutchinson of Ionpure Ltd for discussions on water treatment.

References

ANON. (1972) *Guide to sampling by attributes*, British Standard BS. 6001. British Standards Institution, London.
(1980a) *Guide to data analysis and quality control using cusum charting*, British Standard BS. 5703. British Standards Institution, London.
(1980b) Directive No. 80/778/EEC of 15.7.80 relating to the quality of water intended for human consumption. *Off. J. Eur. Communities* No. 2229, 11.
(1982) *The bacteriological examination of drinking water supplies*, Reports on Public Health and Medical Subjects No. 71. HMSO, London.

(1983) *European Pharmacopoeia*, 2nd edn. Maisonneuve SA, Sainte-Ruffine, France, V.2.1.8–V.2.1.8.10.

(1988) Note for Guidance on the Application of Part 1 of Annex to Directive 75/318/EEC, as amended. In: *Rules governing medicinal products in the EEC*, Vol. III.

(1989) *The rules governing medicinal products in the European Community*, Vol. IV. *Guide to good manufacturing practice for medicinal products*. HMSO, London.

(1990) *CTPA limits and guidelines for microbiological quality management*. Cosmetic, Toiletry and Perfumery Association, London.

(1993a) *British Pharmacopoeia*. HMSO, London.

(1993b) Microbial aspects of pharmaceutical water. *Pharmaceutical Forum 1993*, **19**(5) 6183–6187.

(1995a) *European Pharmacopoeia*, 2nd edn. Maisonneuve SA, Sainte-Ruffine, France, VIII 15.1–15.2

(1995b) *United States Pharmacopoeia XXIII*. US Pharmacopoeial Convention, Rockville, MD.

COWAN, S. T. (1993). *Cowan and Steel's manual for the identification of medical bacteria*. 3rd edn, Cambridge University Press, Cambridge.

CRUIKSHANK, R. (1965) *Medical microbiology*. 11th edn. Livingstone, London, pp. 964–965.

GALLANTREE, I. (1994) Designing a pharmaceutical process water system. Pharmaceutical Manufacturing International. *International Review of Pharmaceutical Technology, Research and Development*, pp. 153–157.

HITOKOTO, H. MOROZUMI, S., WAUKE, T., SAKAI, S., and KURATA, H. (1978) Fungal contamination and mycotoxin detection of powdered herbal drugs. *Appl. Environ. Microbiol.* **36** 252–256.

MILES, A. A., MISRA, S. S., and IRWIN, J. O. (1938) The examination of the bactericidal power of blood. *J. Hyg. (Camb)* **38** 732–749.

PAYNE, D. (1990) Microbial ecology of the production process. In: Denyer, S., and Baird, R. (eds) *Guide to microbiological control in pharmaceutics*. Ellis Horwood, Chichester, pp. 53–67.

(1993) Report of a Joint Pharmaceutical Analysis Group Meeting. *Pharmaceutical Journal*, **251** 30–31.

RUSSELL, A. D., AHONKAI, E., and ROGERS, D. (1979) Microbiological applications of the inactivation of antibiotics and other antimicrobial agents. *J. Appl. Bacteriol.* **46** 207–245.

RUSSELL, M. P., PURDIE, R., GOLDSMITH, J., and PHILLIPS, I. (1984) Computer assisted evaluation of microbiological environmental control data. *J. Parenter. Sci. Technol.* **38** 98–103.

SILLIKER, J. H. (1980a) *Microbial ecology of foods*, Vol. 1, *Factors affecting life and death of micro-organisms*. 2nd edn, International Commission on Microbiological Specifications for Foods, Academic Press, London.

(1980b) *Microbial ecology of foods*, Vol. 2, *Sampling food commodities*, 2nd end, International Commission on Microbiological Specifications for Foods, Academic Press, London.

4

Control of microbial contamination during manufacture

A. CLEGG AND B. F. PERRY

Procter & Gamble Health & Beauty Care, Research and Development Laboratories, Rusham Park, Whitehall Lane, Egham, Surrey TW20 9NW

Summary

(1) Under the Sixth Amendment to the EEC Cosmetics Directive there are now statutory microbiological requirements governing the production of cosmetic products in the UK. Good manufacturing practice (GMP) guidelines are set out in documents prepared by COLIPA and the Council of Europe.

Medicines Control Agency 'Rules and Guidance for Pharmaceutical Manufacturers 1993' set out GMP guidelines covering recent European Commission GMP Directives.

Microbiology guidelines have also been published by the Cosmetic Toiletry and Perfumery Association (CTPA) and by the Cosmetic Toiletry and Fragrances Association (CTFA).

(2) The manufacturing area should be designed so that functions can be physically separated. It should be constructed so that it may easily be cleaned and maintained. Correct ventilation design can prevent accumulation of moisture and the ingress of dust.

(3) Equipment should be selected for the protection that it affords during processing and for ease of dismantling and cleaning. Suitable validated written cleaning and sanitation procedures should be available for each piece of equipment.

(4) All personnel involved in manufacturing operations should be qualified by education and training for the jobs they perform. Suitable, protective clothing, including headwear, should be provided to all staff and should be worn at all times in production areas. Movement of personnel through production areas should be restricted.

(5) Conformance to standards should be checked on a regular basis by means of audits.

4.1 Introduction

Changes in microbiological quality and standards for both the non-sterile pharmaceutical and the cosmetic industry have been by a process of evolution rather than by revolution.

Early attempts to control the quality of products were by the 'end-product'-testing quality control (QC) approach (van Abbé *et al.* 1970). This, of course, only detected contamination after the event, if at all. The statistics of sampling make the detection of low-level or sporadic contamination a hit-or-miss affair.

The pharmaceutical industry led the way in the change from the QC to the quality assurance (QA) approach.

Microbiological QA is a pre-emptive attempt to control microbial contamination by controlling all aspects of manufacturing. This package of control measures is only one aspect of the overall QA approach to manufacturing, usually described by the blanket term good manufacturing practice (GMP).

The principles and guidelines of GMP are stated in the two almost identical EC Directives, 91/356/EEC (Anon. 1991a) for medicinal products for human use and 91/412/EEC (Anon. 1991b) for veterinary medicinal products. In the UK these Directives have been implemented by standard provisions and undertakings incorporated in regulations made under the Medicines act (Anon. 1968). Whereas before 1992 there was no specific legal requirement for pharmaceuticals to comply with GMP, compliance with the principles and guidelines of GMP is now a statutory requirement. The European Community Guide to GMP (Anon. 1992a) provides detailed guidance which interprets and expands on the statutory Principles and Guidelines.

The UK first produced a national guide to GMP (known affectionately as the 'Orange Guide' (Anon. 1977, 1983) from the colour of its cover) in 1971 (Anon. 1971). Two further editions were produced in 1977 and 1983 (Anon. 1977, 1983). The EC Guide, issued first in 1989 and reprinted with minor corrections and additional annexes in 1992, supersedes this and all other national guides of EC Member States (Anon. 1992a). Further annexes (numbers 13 and 14) were adopted in 1993 (Anon. 1993a). The Pharmaceutical Inspection Convention has also adopted the text of the EC Guide ensuring harmonization of guidelines throughout Europe and beyond.

In the cosmetics industry, article 7C of the Sixth Amendment to the EEC Cosmetics Directive (Anon. 1993b) states that the method of manufacture has to comply with the good manufacturing practice (GMP) that will be laid down in a future EC regulation, or as a first step, in a national one.

Unlike the case for drug or veterinary medicinal producers (Anon. 1991a, 1991b) no specific official guidelines exist for cosmetic manufacturers. The 'Advisory Notes to Manufacturers of Cosmetics and Toiletries' were published by the European Cosmetic Toiletry and Perfumery Association (COLIPA) in 1988, but for voluntary and internal use only (Anon. 1988).

The French Health Office has recently adopted a text specifically related to GMP prepared in collaboration with the French cosmetic trade association (Anon. 1992b). This guideline has been drawn up to conform with international norm ISO 9004 'Quality Management and Quality System Elements' (Anon. 1994a). These guidelines are intended to give aid on all aspects of cosmetic product manufacturing and describe the installation of quality systems and their impact on each stage of cosmetic manufacture. Both COLIPA and the Council of Europe have produced GMP guidelines (Anon. 1994b, 1995a). COLIPA has also produced guidelines on the content of product information files, as required by the above EEC Cosmetics Directive (Anon. 1995b).

The UK cosmetic trade association, the CTPA, recommend a total quality approach to microbiological quality assurance, termed Microbial Quality Management and published its MQM limits and guidelines in 1990 (Anon. 1990a). These guidelines are currently being updated and publication is expected in 1996 (Anon. 1996). The CTPA has also published guidelines to cosmetic manufacturers who wish to conform to BS 5750 Part 2 (ISO 9004) (Anon. 1990b). Microbiological guidelines have also been published by the US cosmetic trade association, CTFA (Anon. 1993c). Other useful guidelines can be obtained from Yablonski (1978), Russell (1983a, 1983b, 1983c), Diel (1992) and Warwick (1993) and found within books by Denyer and Baird (1990) and Orth (1993). This chapter will attempt to outline the salient points of microbiological in-process control.

4.2 Sources of microbial contamination

In order to gain a proper understanding of GMP, it is necessary to break down the various stages of manufacture and, for each stage,

(1)   to identify the various sources of contamination and the nature of the contamination likely to be present and

(2)   to devise suitable methods by which the contamination may be controlled and prevented from entering the product.

If we examine these stages we find that, during manufacture, the bulk product may become contaminated from the raw materials, the manufacturing equipment and environment and the manufacturing personnel. When the bulk product is filled into final containers, it may be contaminated from the filling equipment and environment, the container or the personnel involved in the filling operation. Original pack products may be further contaminated as a result of dilution (e.g. from concentrates) and repacking by the consumer from refill packs (Perry 1994).

If we look in more detail at these various sources of contamination, we find that probably the most important of these is the raw materials. This source of contamination has already been discussed (see chapter 3). Unless great care is taken, the manufacturing and filling plant will most certainly act as a contamination source since it comes into direct contact with the product. Between preparation of product batches, growth of microbial contaminants can easily occur in 'dead spaces' (particularly joints and valves) where product residues readily accumulate. Experience has shown that, once established, such contamination can be very persistent and difficult to eliminate. Although contamination of the manufacturing environment may be considered as less important since it is not in direct contact with the product, contamination of products from this source has been demonstrated in situations where environmental control is not properly implemented (Baird 1977). For dry surfaces such as floors and walls, contamination comprises mainly Gram-positive cocci, aerobic Gram-positive rods and fungal spores. In general, Gram-negative bacteria are more susceptible to the lethal effects of drying although small numbers of these organisms may persist on dry surfaces for quite substantial periods of time. Damage to hard surfaces is known to encourage the accumulation of contaminants. In damper wet areas such as sinks and drains, particular where there is stagnant water, nutritionally non-exacting Gram-negative species such as *Pseudomonas* spp. and *Acinetobacter* spp. will not only survive but can readily proliferate to form reservoirs of free-living microbial contamination. Other surfaces where dampness is not controlled will also encourage the survival of both Gram-negative and Gram-positive species. Contamination may also originate from the air in the manufacturing environment. Air-borne organisms rarely occur in isolation and are usually associated with dust and skin scales. The organisms found are mainly mould and bacterial spores, and skin cocci. Probably the greatest potential sources of contamination is that from the operators as this is the least controllable. An occupied room will always contain skin fragments in the dust and air. It is estimated that during walking, the loss of skin scales is about $10^4\,min^{-1}$ and, although this is reduced when a person is stationary, the loss rate is still considerable. It is known that a large proportion of these skin scales bear bacteria. The normal skin flora consists mainly of non-pathogenic micrococci, diphtheroids and staphylococci. It is estimated that up to 30% of the population may act as permanent or transient carriers of *Staphylococcus aureus*. In addition to the resident skin flora, other micro-organisms may be carried transiently on the skin and may be shed into the product either via skin scales or by direct contact.

In the following sections the various methods which are currently used to prevent transfer of contamination from these various sources to the product are considered.

4.3 The manufacturing environment

For the purposes of this chapter, the environment refers to those factors, other than process equipment and people, which directly influence the microbiological quality of a finished product. These include the design of buildings and processes, the choice of construction materials and surface finishes, the provision of ancillary services such as lighting, ventilation, power, etc. and the exclusion of vermin.

4.3.1 *Design and layout*

Early QA and microbiological input to the design of manufacturing processes and facilities, will make the implementation and practice of GMP a much easier process. The primary concern, at the initial design stages, is to establish a unidirectional flow of materials. An idealized flow diagram is given in Fig. 4.1. This ensures that there can be no cross-flow of clean and contaminated materials and maintains the 'integrity' of the manufacturing operations.

Raw materials should enter the manufacturing area through a defined dispensing area. Materials taken into this area should be free of any extraneous dirt or moisture. For example, outer cartons and packaging materials, particularly straw, wood shavings or sawdust, should be removed before primary containers are taken into the manufacturing area.

Drums should be free of dust or moisture and, if necessary, should be washed down or disinfected and dried before taken into stock. Pallets used in raw materials warehouses should never be taken into the dispensary. If necessary, metal or plastic pallets should be provided for storage and transportation of dispensed materials. These should never leave the manufacturing area.

Similar restrictions should be applied to the movement of people through manufacturing areas. All unnecessary traffic should be banned. Personnel access to manufacturing areas should be through a changing area. Protective clothing, including head coverings, should be provided and worn. This clothing should never be worn outside the manufacturing areas and street clothing should never be allowed in the manufacturing area.

The actual level of segregation of operations will depend upon the type of operation and the types of products being produced. In a pharmaceutical plant, at one extreme, the areas will be physically separate and may be sealed units with access through air locks. In

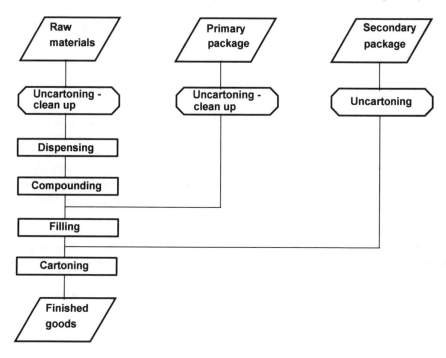

Figure 4.1   Idealized flow diagram for manufacturing processes

other plants the areas may be defined simply by lines on the floor, but the basic principles are the same in each case.

4.3.2   Principles of construction

Walls, floors and ceilings should be constructed of materials that are smooth, non-porous and non-shedding. If this is not possible, then the surfaces should be suitably sealed. Particular attention should be given to the choice of flooring. It must be able to withstand the projected loading without breaking up. A cracked floor cannot be properly cleaned.

All floors should slope towards a trapped and vented drain (one drain per 40 m^{-2}). The surface should be smooth, but not slippery. The junctions of walls and floor should be rounded and the flooring should be continued a little way up the wall to form a cove. This avoids corners where dirt can accumulate.

Any windows should be non-opening and flush-fitting. Window ledges attract dust and should be avoided. Any projecting corners, particularly around doorways, should be protected against damage by means of metal plates. Damaged walls absorb water and release dust into the air. Any such plates should be set flush into the wall or the edges should be sealed to prevent ingress of dirt and water.

Light fittings should be flush-fitting, with no ledges to trap dust.

4.3.3   Provision of services

When providing services to any manufacturing area, the cardinal rule is to avoid dust traps. The usual routes for electrical supplies, air and steam lines and product transfer lines are the overhead route or via in-wall trunking. As a general principle, surface wall mountings should be avoided. They trap dirt and make cleaning difficult. Overhead services should be run in the ceiling void above a suspended ceiling. Services can be kept conveniently out of the manufacturing area and structural roof-members covered up whilst still maintaining easy access.

Ventilation is probably the most difficult service to get right. Air flows through manufacturing areas should be kept to an absolute minimum. This almost always entails the provision of local supply and extraction. Air supplies should be filtered to remove dust and screened to prevent the entry of insects. Air-conditioning systems can be a particular problem. Evaporation coolers and humidifiers can be very efficient breeding grounds for bacteria. If they must be used, then a bacterial filter should be placed downstream and a regular cleaning and sanitation routine implemented and recorded.

If parts of the manufacturing process generate large volumes of steam or dust, then the ventilation system can become overloaded. In this case, local extraction must be provided to prevent cross-contamination with dust or condensation. The filtered air supply must be able to provide the make-up air volume, or contaminated air will be drawn directly into the building. For this reason it is better to have continuous, rather than intermittent, local extraction.

4.3.4 Hygiene policy

Cleaning regimes for the manufacturing area should be designed to produce as little dust as possible. It is better to have major clean-downs when the plant is not in production, at night or between shifts. Floors should be washed daily. During production any spillage of product or accumulation of moisture should be cleaned up as soon as possible and, preferably, immediately. A schedule should be developed for the periodic washing of walls and ceilings and a responsible person should be nominated to see that the programme is carried out.

Cleaning equipment should not, itself, be allowed to become a source of contamination. Mops, buckets, squeegees and brushes should be kept in a sanitary condition, which means clean, dry and stored off the floor. Floor-washing machines, particularly those with integral tanks for detergents, can be dangerous. Bacteria can breed in the solutions or the delivery system and the machines then become disseminators of contamination. Such machines should themselves be designed for easy cleaning and sanitation. Cleaning equipment should be stored in a designated wash area, which should itself be kept clean and tidy.

Toilets and hand-washing facilities should be provided outside the production area. Return to the production area should only be possible by passing a hand basin.

People can be one of the greatest sources of contamination in the plant. All employees should be provided with clean, protective clothing, including head coverings, as often as necessary. All personnel should be encouraged to report minor ailments that may serve as a potential source of contamination. They, and anyone having exposed sores, cuts or abrasions, should be assigned to duties that do not involve exposure to products. The wearing of jewellery, badges, hair clips, etc. should not be allowed.

4.3.5 Control of vermin

A vermin control policy should be in place. This should include physical systems for the exclusion of insects, rodents and birds. A pest control officer should be appointed. He or she may be a suitably qualified employee, or a consultant from an outside contractor.

Some simple rules can help to reduce the chances of infestation. Doors should be kept closed at all times. Loading bay doors should be provided with seals so that loading operations are carried out under cover. Eating and drinking, except in the designated place, should be strictly forbidden. Food attracts vermin.

Rubbish and excessive quantities of packaging materials or finished stock should not be allowed to accumulate in the production area.

4.3.6 Environmental monitoring

Environmental conditions may directly or indirectly affect the quality of the finished product.

An environmental monitoring programme has to be properly constituted to be of any practical use. It is left to the discretion of the manufacturer to determine what type of programme would be most suitable for his or her own facility, since each plant is unique. External standards for the microbiological quality of air and surfaces do not exist for non-sterile production. Most manufacturers, therefore, develop their own 'in-house' standards. Such standards must be based on knowledge of the normal background levels of contamination. These must be determined at different times of the day over an extended period, preferably to include all the seasons, before a realistic picture can be determined and standards set. Regardless of the approach used, a manufacturer should establish guideline and action levels for all the areas tested, once a normal range of recovery has been established. If any results exceed the action levels, the group responsible for cleaning and sanitizing should be immediately informed and corrective action should be taken. Exceeding the manufacturer's guidelines should precipitate further testing, including the finished product.

The quality of the manufacturing environment will be influenced by many factors. Understanding these influences is an essential part of an environmental control programme. Influences on the environment can be divided into four basic categories: buildings, equipment, personnel, and sanitization and housekeeping. These are discussed elsewhere in this chapter.

The microbiological quality of physical surfaces within a manufacturing environment can be defined as either

direct contact surfaces: e.g.	processing equipment filling equipment pumps valves
or non-direct contact surfaces: e.g.	walls, floors, ceilings overhead lighting or walkways support beams pallets, forklift trucks.

The type and frequency of microbiological monitoring of physical surfaces will depend on the susceptibility of the product to microbial contamination, the scale of manufacturing and the type of plant and process. Those physical surfaces which come in direct contact with the product would normally be more frequently monitored than those not in direct contact with the product. Direct contact surfaces should be examined for the presence of micro-organisms that are known to cause harm to the consumer or product spoilage. Indirect surfaces should be monitored to determine background levels of micro-organisms which are intrinsic to the manufacturing environment. Any increase from these predetermined levels may be considered an indication of a potential microbiological problem that may affect product quality and may necessitate a revision of the frequency of testing and sanitization.

The most commonly used methods include the use of swabs, contacts plates and the collection of rinse water. The relative advantages and disadvantages of these approaches have been reviewed by the CFTA (Anon. 1993c). Personnel involved in environmental monitoring should be properly trained according to a written procedure and should establish a properly documented programme. In general, all processing and filling equipment, valves, working surfaces etc. should be monitored on a defined and periodic basis. Transfer lines should be taken apart, if necessary, and rinsed thoroughly with an appropriate rinse solution of known microbiological quality.

A schedule should be prepared and followed for the routine monitoring of the air in each designated area. The frequency of monitoring should be determined by the type of activities in the area. Again, frequency and standards should be established by 'in-house' needs. The most commonly used methods, many of which are available commercially, include:

sedimentation:	settling agar plates
filtration:	membrane filter
impingement:	all glass impinger
impaction:	Andersen-sieve, slit-to-agar
	centrifugal air sampler

Again, the relative advantages and disadvantages of these methods have been reviewed by the CTFA (Anon. 1993c). Detailed guidance on sampling methods can also be found in a Parenteral Society monograph (1989) and Baird (1990).

Environmental monitoring serves a useful purpose as a control and feedback mechanism to monitor and validate cleaning and sanitization procedures (see section 4.4.4).

4.4 Manufacturing equipment

One of the greatest contributions the microbiologist can make to hygienic manufacture is the advice given on choice of equipment. Cooperation among purchasing, engineering, microbiology and quality assurance can help to prevent many future problems. For example, ease of cleaning and sanitation should be given serious consideration when purchasing any new equipment.

4.4.1 *Compounding and bulk storage equipment*

Compounding and bulk storage equipment must be designed with cleaning in mind. Materials of construction have to be able to withstand routine cleaning and sanitation. The best material, without doubt, is stainless steel, 316L grade or better for all product-contact surfaces.

A stylized compounding facility is shown in Fig. 4.2. The points requiring particular attention to be paid to hygienic design are indicated. Generally, all internal joints on vessels should be argon-welded and ground smooth. Tank corners should be rounded and drain valves should be located at the lowest point in the tank. Tank bottoms should be sloped so that they are self-draining.

All tanks should have close-fitting lids or be totally sealed with the only access via a manway. Side entry manways should be avoided as they require gaskets that come into contact with product. Similarly, side and bottom entry mixers or impellers should be avoided, since they require glands. These can become contaminated with product and, thus, become a source of contamination. Top entry mixers should be mounted so that oil or dirt cannot fall from them into the product being mixed. Portable mixers are a particular problem in this context.

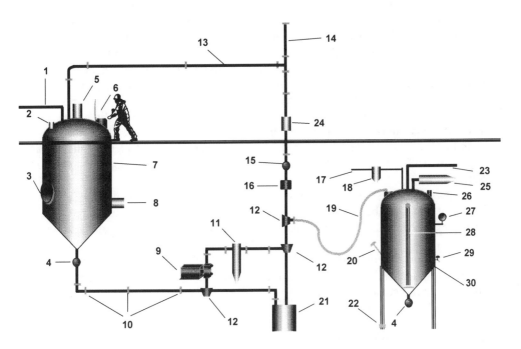

Figure 4.2 Items requiring attention to hygienic design: 1, product-water inlet; 2, anti-vacuum valve; 3, side-entry manway; 4, outlet valve; 5, agitator; 6, top-entry manway; 7, manufacturing vessel; 8, side-entry mixer drives, glands and seals; 9, pump; 10, pipeline connectors and joints; 11, in-line filter; 12, three-way valves; 13 horizontal pipe run; 14, dead leg; 15, valve; 16, flow meter; 17, air-line; 18, air-filter; 19, flexible hose; 20, thermometer pocket; 21, expansion vessel; 22, load-cell; 23, CIP system; 24, non-return valve; 25, filtered air-bleed; 26, pressure-relief valve; 27, temperature or pressure gauge; 28, sight-glass; 29, sample cock; 30, storage vessel

Ancillary control equipment, such as flow meters, pressure gauges, thermometer pockets and level gauges, needs careful attention if it is not to become a collecting point for old product and cleaning solution. Externally mounted level glasses, particularly if they are open at the top, are impossible to clean properly and provide a rapid and easy means of entry for bacteria.

Air bleeds or vents must be protected by a bacteriological air filter. More importantly, there must be a written procedure for checking and changing the filter and responsibility for carrying it out must be clearly defined.

Transfer lines and pumps that cannot be adequately cleaned and sterilized are frequent sources of contamination. Pipelines should be constructed of 316L grade stainless steel and should preferably have welded joints, ground smooth. Sanitary fittings are an acceptable alternative, but rubber gaskets, flanges or threaded joints must not be used. Teflon or Viton are acceptable materials for gaskets. All pipelines should be sloped to a venting point and should be adequately supported. A horizontal line, or worse, one that sags in the middle allows a build-up of stagnant product. Dead ends and bypasses, which are infrequently used, can also become reservoirs of contamination and should be eliminated. Lines should run straight, with no right-angled or vertical bends.

Valves require particular attention. The common types used in pharmaceutical and cosmetic manufacture are plug, ball, butterfly and diaphragm valves. Of these the butterfly type is probably the most acceptable. Some butterfly valves can be cleaned *in situ*, though they must be stripped and examined for damage or accumulation of product residue at regular intervals. Ball valves and plug valves both have dead spaces in which product can be trapped. Dairy grade valves of both types, however, can be easily dismantled and re-assembled for cleaning purposes. If they are used, they should be stripped and examined as part of the cleaning procedure. Needle valves are impossible to clean properly and should be avoided. Diaphragm valves are subject to 'pin holing' of the diaphragm and can become sources of contamination.

In-line filters have been implicated in many contamination incidents. Filter presses require particular attention, but even cartridge filters require regular cleaning and maintenance.

Sanitary pump design and choice of the correct pump for the job in hand are extremely important. The number of designs, types and variants is legion, but the same basic principles apply to them all. They must be external to the system and should be accessible for dismantling and cleaning. They should have no dead volumes and should be completely self-emptying. Glands and drives should be checked frequently to ensure that they are not worn and leaking. Sanitary lobe pumps and double diaphragm pumps are probably the most acceptable designs.

4.4.2 *Filling equipment*

The types of and variations on filling equipment are numerous, be they vacuum, gravity or displacement; single, double or multi-nozzle; in-line or rotary head. The principles of sanitary design can, however, be applied equally to all of them. A stylized filling line is shown in Fig. 4.3; the points requiring sanitary design are indicated.

All filling equipment should be capable of easy stripping for cleaning and maintenance. If it takes five hours to strip and clean a machine it is less likely to be done.

Line holding or break tanks require careful design. They are a point of stagnation for product and should be kept as small as possible or, preferably, avoided altogether. Ball

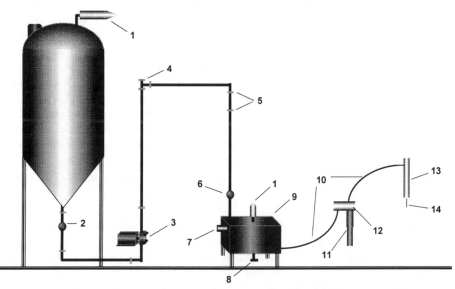

Figure 4.3 Filling line showing points needing hygienic design: 1, filtered air-bleeds; 2, outlet valve; 3, product-transfer pump; 4, pipe bend; 5, pipe connectors and joints; 6, valve; 7, level switch; 8, drain valve; 9, holding tank; 10, flexible hoses; 11, piston; 12, dosing pump; 13, filling head; 14, filling nozzle

valves should be avoided; maintenance of the correct product level should be via a sanitary level-switch operating a pump or, on a gravity system, a pneumatic valve. The valve should be easily stripped for cleaning. Bypasses and dead ends should be ruthlessly eliminated.

Conveyors, turntables or other areas where open bottles are exposed to the air should be suitably protected against dust or, preferably, fully enclosed.

Air lines or bleeds must be fitted with bacteriological filters. These must be inspected and changed according to a written schedule and records kept. Bottle blowers should be included in the schedule.

4.4.3 *Water treatment systems*

Water systems are particularly susceptible to microbial contamination and must be designed to prevent microbes from proliferating and thence being introduced into products. There should be systems in place to remove suspended matter, silicates, organic materials, chlorine, and colloidal material from the raw water before it enters the treatment system.

Modern water treatment systems have a multi-media filter specifically designed according to the quality of the incoming water. Such filters combine depth filtration with activated charcoal and coarse ion-exchange resins. The routine microbial monitoring programme should provide information for the operator to determine whether chemical disinfection of the demineralization equipment is required. The suitability of any such disinfecting agent should be confirmed with the resin supplier. Where required, anti-microbial treatment systems must be incorporated.

Ozonization, at 0.1ppm maximum, is an effective disinfectant system, although with relatively high maintenance and operation costs. Appropriate ultraviolet (UV) lights must be installed at points of use to remove the ozone. Water must be recirculated at an adequate velocity (minimum of 1 m/s) to ensure even distribution of the ozone.

UV irradiation at 254 nm is effective in clear, colourless water. UV transmission can be reduced by particulate matter, organic material and some pigments. UV lamps generate large amounts of heat in use and can only operate when water is flowing. They take up to 30 seconds to reach operating wavelength after being switched on and any water flowing through in that time will not be fully treated. It takes a set time for UV to disable or kill microbes and, in a moving system, there may not be sufficient dwell time in the lamp chamber to ensure effectiveness. These problems can be resolved by moving the water through the UV lamps in a continuous recirculation system.

Heat can be used to prevent microbial growth in water systems. The water must be maintained above 85°C at all times throughout the system. In order to ensure even distribution of the heat, recirculation is required. Heated systems have high operating charges and high installation costs for insulation. Cooling systems would be needed for processes requiring water at less than 85°C.

Bacterial filters should not be used as the sole mechanism for microbial control due to the potential for bacteria to grow on the filter. Filters used at point of use should either be capable of resterilization or be thrown away after each use. Filters can become blocked after a period of usage and may restrict water flow. It is therefore important to monitor the pressure differential across the filter.

In order to maintain the microbial integrity of the treated water, it must continually flow through the anti-microbial system. This can be achieved by keeping the treated water moving in a continuous loop. Water should be distributed in a pipework loop separate from that of the supply system. When there is no water usage, water flow in both loops can then be maintained. This is frequently referred as a 'figure-of-eight' system. Fig. 4.4 illustrates a simple 'figure-of-eight' water system.

Water flowing through a pipe creates a 'viscous' boundary layer at the pipe surface due to the effect of drag caused by water surface tension. Any microbes falling into this layer will be in a relatively calm zone and will proliferate to form a film of cells across the inner surface. This biofilm will eventually grow into the flowing water and clumps will be broken away. If these clumps are large, anti-microbial systems may not affect those cells in the centre of the clump.

Flow rates should be high enough to ensure that the viscous layer is not more than 5 μm. This should limit any biofilm to no more than five cells in depth and any clumps broken off will be small enough for the anti-microbial treatment to handle. The minimum flow rate to achieve this condition depends on the diameter of the pipe.

4.4.4   Sanitation

There should be a clear, written sanitation policy, expressed unambiguously, in plain language.

For each piece of equipment, there should be a sanitation schedule and written method available. Sanitation should be recorded and signed off by the person performing the task. The method must be written in sufficient detail to allow any operator to carry it out, though certain individuals should be made responsible for cleaning and sanitation. The

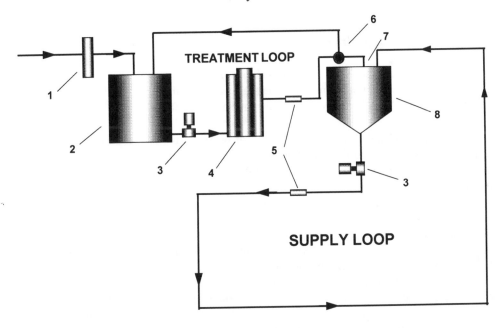

Figure 4.4 Figure-of-eight water system: 1, multi-media filter; 2, city water storage tank; 3, pumps; 4, demineralizer; 5, antimicrobial treatment; 6, return valve; 7, sanitary level switch; 8, treated water storage tank

first stage of any procedure is thorough cleaning. Equipment that can be stripped and cleaned should be. Residual product may inactive sanitizing agents or protect bacteria from their action.

Cleaning in place (CIP) is a semi- or fully-automated, self-contained system for cleaning and sanitizing equipment without disassembly. Each system is unique and if it is to work well it should be properly designed, evaluated and controlled. CIP should not be considered as an add-on to existing plant or procedures. In a CIP system, hot detergent solutions are delivered, usually under high pressure, through nozzles or spray heads, so that all parts of the equipment are cleaned. Factors to consider when using CIP are: detergent/sanitizer type; detergent/sanitizer concentration; temperature and design of equipment.

The greatest danger from CIP systems is that they can, themselves, become sources of contamination. They are often heated by steam injection into a steam and water mixer and the water is not usually pre-treated. Bacterial numbers can increase alarmingly between use of the system. For this reason, the CIP system itself has to be designed for ease of cleaning and sanitation.

Equipment design should consider (Anon. 1993c):

- Velocity of detergent or sanitizing solution.
  — Velocity should be approximately four times that of product flow.
  — For lower velocity, more attention should be focused on detergent type, its use, concentration and temperature.

- Re-use of detergent/sanitizer solution.
  — Requires thorough rinsing of product residue prior to application of detergent/sanitizer.
  — Fresh detergent should be added to maintain proper concentration prior to each use.
  — Detergent/sanitizer solution should be replaced as required.
- Programmed automatic control system for time and temperature.

Time spent in cleaning is not wasted time. It increases the efficacy of the sanitation steps and can even reduce the necessary contact times. After cleaning, sanitation can be carried out. The most effective sanitizer is moist heat, but chemical methods, if properly used, can be effective.

When steam is used, the minimum temperature attained in any part of the system should be 82°C. Timing is started when the steam exiting from the system at the farthest point reaches this temperature. The minimum contact time is 30 minutes.

The most effective chemical sanitizers are chlorine-releasing compounds or iodophors. Other agents such as quaternary ammonium compounds or amphoterics can be used. Chlorine compounds may be used at a level of 200 to 250 mg l^{-1} (ppm) of available chlorine in static systems and will achieve sanitation in 30 minutes. If circulation is introduced, it may be possible to use as little as 50 mg l^{-1}, since the relative efficacy is increased. Iodophors are as effective as chlorine compounds, but are more difficult to rinse from the system. They are not, therefore, widely used.

If non-halogen sanitizers are used, then a system of rotation of sanitizing agents must be introduced, in order to prevent the development of resistant strains of bacteria.

All sanitizers must be rinsed thoroughly from the system before it is used. For this purpose only treated process water should be employed or contamination may be re-introduced. In systems where product is used for rinsing, the sanitizer must be compatible with the product. Whatever cleaning and sanitation process is chosen, it must be thoroughly validated for each piece of equipment and for each product run on that equipment.

4.5 Personnel

It is important that the implementation of a GMP programme should have the input of a qualified microbiologist. He or she does not have to be a direct employee of the company, but may be an external consultant. In this case, consultation should be on a regular basis and an internal manager should be made responsible for the implementation of the programme.

Management should be seen to be committed to GMP or the programme will lack credibility with the employees. It is no use declaring a micro-biological control policy and then, by the actions of management, demonstrating that profit will always override quality concerns.

It is important, also, that all staff at all levels should feel involved in, and responsible for, microbiological quality. This is achieved by ensuring that people are properly trained in hygienic manufacture, including personal hygiene, from the outset and that this training is regularly reinforced. Training records should be kept and regularly updated.

All employees should be encouraged to bring quality matters to the attention of management.

4.6   Validation, conformance and audit

4.6.1   *Monitoring of production*

In a GMP environment, finished product testing is not the major means of controlling the microbiological quality of products. Its objectives should be to check on the overall level of GMP and to detect major breakdowns in product quality. It cannot detect low level or sporadic contamination with any degree of certainty because of the inherent statistics of sampling. Only when combined with a GMP programme can it give the required level of assurance.

In terms of GMP management, the days of accepting with blind faith a retrospective control maxim for release of product have now passed. Backward management must now be replaced by a prospective or forward control strategy. Such an approach had been advocated in the food industry since about 1920 (Mossel *et al.* 1994). However, until the late 1960s this advice had largely been ignored and had therefore, hardly affected consumer protection practices until Bauman heralded his Hazard Analysis of Critical Control Points (HACCP) maxim (Bauman 1974). This proved to be readily acceptable in the food industry. In contrast, the earlier warning of Dack (1956) has barely been heeded: unless HACCP was *extended longitudinally*, from raw materials to consumption, it might well remain fruitless. Interestingly Bauman's critical points and their rectification had previously been identified much earlier by Wilson (1933) who recommended a polypragmatic management policy, since known as the Wilson Triad.

Thus intervention has to start by quality assurance of raw materials and in the course of production hygienic measures must be taken to eliminate 'critical' sites and procedures and, for foods, in many instances a terminal decontamination step proves indispensable. Basically, this is a rational process of estimating the risk associated with each step in the manufacturing process and supply to the consumer. It involves consideration of raw materials, the processing steps and how the product is delivered to the consumer and potentially how it is used by the consumer. Hazard analysis is, therefore, an integral part of any validation exercise. It is designed to identify the points at which a process can go wrong, the critical control points. By monitoring these points, the whole process can be kept under control.

The total quality concept thus emerging has for some time now been designated as: Longitudinal Integrated Safety Assurance (LISA).

Adoption of a Wilson–Dack–Bauman HACCP–LISA strategy to the production of cosmetics, toiletries and pharmaceuticals entails the introduction of efficacious management approaches which will lead to dramatically reduced spatial and temporal fluctuations in the microbiological profile of these products, henceforth produced. Consequently, once control is achieved, analytical data obtained on samples no longer lack predictive significance. Elevated colony counts will hence point to LISA being temporarily out of control, in spite of critical situations having been effectively eliminated. This allows early identification of trouble-spots not previously anticipated, and, most importantly, their immediate rectification.

The basic components of this strategy are:

(1)   the biological critical points which, if not maintained within certain parameters can result in the manufacture of product of unacceptable microbiological quality, and

(2)   the physical hazard control point which is an element of process equipment or environment of the process that, if not properly maintained and monitored can result in the introduction of hazardous foreign materials into the product.

For a more detailed discussion of HACCP and LISA, see Mossel *et al.* (1995).

In recent years increasing emphasis has been placed upon the importance of validation in the manufacture of high-quality products. Validation is the provision of documentary evidence that a process is able to make a product both reliably and reproducibly. There should be a clear written validation master plan, approved at high company level, outlining company philosophy and requirements. Each validation requires a detailed protocol and full report, including any deviations, approved by qualified personnel. Before process validation, equipment needs to be qualified during installation and operation. Validation is not only a GMP requirement, but a key part of good formulation development and manufacturing risk management.

Microbiological sampling at various points in the production process can be an effective way of validating the hygienic procedures used. A 'bracketing' technique – analysis of samples before and after a particular process step (e.g. before and after a storage tank, line pump, homogenizer, filler etc.) – can be used to demonstrate whether that step may add to the microbial load of the product stream. Data can thus be developed, for example, on the effectiveness of a cleaning procedure or to identify critical control points in a HACCP scheme.

The efficacy of cleaning procedures can be validated by physically stripping down equipment and looking for product residue and by swabbing for the presence of microorganisms.

Thus to summarize, validation is a formal programme of testing designed to show that a product can be reliably manufactured, using a given process. It has to generate, record and interpret sufficient information to reassure that the process can be controlled and to identify the critical control points. Continued monitoring of the Critical Control Points ensures that the process remains under control.

The development of such a programme is, necessarily, a team performance.

4.6.2   Conformance and audit

Conformance of product specification, standard operating procedures, cleaning procedures, etc. must be formally checked on a regular basis. This is achieved by means of a microbiological audit (Anon. 1992c). Such an audit can take one of two forms, the external audit and the internal or self-audit.

The external audit is carried out by a person or persons not directly involved with production. They may be representatives of R&D or QA.

The self-audit is a team audit, involving representatives of all the major functions, including microbiology. This kind of audit usually concentrates on one department at a time, according to a predetermined schedule. It can profitably be made part of the Manufacturing Self-Inspection (Improvement) Programme (MSIP).

In practice, there is no difference in the way in which either audit is carried out. Audits should not be looked on as fault-finding exercises, but as tools to help the continued improvement of GMP. It can be helpful to have checklists to ensure that points are not overlooked, but over-reliance on such aids can result in a 'blinkered' approach.

At the end of each audit, a report is prepared. This report should be a joint effort between the audit team and the people being audited. It should contain a plan of actions to be taken to improve any deficiencies identified and should set follow-up dates to review implementation. At these dates, follow-up reports should be written and any deficiencies should be corrected as quickly as possible.

References

ANON. (1968) *Medicines Act.* HMSO, London.

(1971) *Guide to good pharmaceutical manufacturing practice.* HMSO, London.

(1977) *Guide to good pharmaceutical manufacturing practice.* HMSO, London.

(1983) *Guide to good pharmaceutical manufacturing practice.* HMSO, London.

(1988) *Principles of good manufacturing practice in the toiletry and cosmetic industry.* European Trade Federation of the Toiletries and Cosmetics Industry (COLIPA), Brussels.

(1990a) *MQM microbial quality management. CTPA limits and guidelines.* Cosmetic Toiletry and Perfumery Association UK, London.

(1990b) BS 5750: Part 2: 1987 (ISO 9004). *Guidelines for use by the cosmetic industry.* Cosmetic Toiletry and Perfumery Association UK, London.

(1991a) Commission Directive 91/356/EEC of 13 June 1991 laying down the principles and guidelines of good manufacturing practice for medicinal products for human use. *Off. J. Eur. Comm.* L 193/30 of 17.7.1991.

(1991b) Commission Directive 91/412/EEC of 23 July 1991 laying down the principles and guidelines of good manufacturing practice for veterinary medicinal products. *Off. J. Eur. Comm.* L 228/70 of 17.8.1991.

(1992a) Good manufacturing practice for medicinal products. In: *The rules governing products in the European Community*, Vol. IV. Commission of the European Communities, Brussels & Luxembourg.

(1992b) *Good manufacturing practices.* Federation des Industries de la Parfumerie, Paris.

(1992c) *CTPA GMP Audit. Audit guidelines for the cosmetic industry.* Cosmetic Toiletry and Perfumery Association UK, London.

(1993a) *Medicines Control Agency Rules and guidance for pharmaceutical manufacturers.* HMSO, London.

(1993b) Council Directive 93/35/EEC of 14 June 1993 amending for the sixth time Directive 76/768/EEC on the approximation of the laws of the Member States relating to cosmetic products. *Off. J Eur. Comm.* L 151/32 of 23.6.93.

(1993c) *CTFA Microbiology guidelines.* Cosmetic Toiletry and Fragrance Association US, Washington, DC.

(1994a) BS EN ISO 9004: *Quality management and quality system elements—guidelines.* British Standards Institution, London.

(1994b) *Cosmetic good manufacturing practices. Guidelines for the manufacturer of cosmetic products.* The European Cosmetic Toiletry and Perfumery Association (COLIPA), Brussels.

(1995c) *Proposed guidelines of good manufacturing practices of cosmetic products*, Revision 3. Council of Europe. Brussels.

(1995b) Product information [September 1993]. Guidelines to the industry based on the Sixth Amendement to EEC Cosmetics Directive. COLIPA, Brussels.

(1996) MQM microbial quality management. CTPA limits and guidelines. Cosmetic Toiletry and Perfumery Association UK, London (in press).

BAIRD, R. M. (1977) Microbial contamination of cosmetic products. *J. Soc. Cosmet. Chem.* 28 17–20.

(1990) Monitoring microbiological quality: conventional testing methods. In: Denyer, S. and

Baird, R. (eds), *Guide to microbiological control in pharmaceuticals*. Ellis Horwood, Chichester, pp. 125–145.

BAUMAN, H. E. (1974) The HACCP concept and microbiological hazard categories. *Food Technol.* **28**(9), 30–34, 74.

DACK (1956) Evaluation of microbiological standards for foods. *Food Technol.* **10** 507–509.DENYER, S., and BAIRD, R. (eds) (1990) *Guide to microbiological control in pharmaceuticals*. Ellis Horwood, Chichester.

DIEL, K.-H. (1992) The key to microbiological quality assurance. *Seife Oele Fette Wachse* **118** 136–146.

MOSSEL, D. A. A., STRUIJK, C. B., PASTONI, F., MARENGO, C. M. L., and FEROLI, M. (1994) Ecological essentials of ensuring and auditing the microbiological integrity of colonization – prone foods and their dissemination through modernized internationally oriented postgraduate education. In: Proceedings Third International Symposium Microbiology of Food and Cosmetics in Europe 1993: Yesterday's Projects and Today's Reality. Marengo, G. and Pastoni, F. (eds). Joint Research Centre EUR 15601 EN Luxembourg Europena Commission, pp. 7–30.

MOSSEL, D. A. A., CORRY, J. E. L., BAIRD, R. M., and STRUIJK, C. B. (1995) *Essentials of the microbiology of foods*. John Wiley, Chichester.

ORTH, D. S. (1993) *Handbook of cosmetic microbiology*. Marcel Dekker, New York.

PARENTERAL SOCIETY (1989) Environmental contamination control practice. Technical Monograph No. 2.

PERRY, B. F. (1994) Adverse health consequences associated with the microbiological contamination of cosmetics: role of microbial quality management strategy in reducing risk. In: Proceedings Third International Symposium Microbiology of Food and Cosmetics in Europe 1993: Yesterday's Projects and Today's Reality. Marengo, G. and Pastoni, F. (eds). Joint Research Centre EUR 15601 EN Luxembourg European Commission, pp. 163–175.

RUSSELL, M. P. (1983a) Microbiological aspects of control in manufacturing. Part 1. *Soap Perfum. Cosmet.* **May** 234–236.

(1983b) Microbiological aspects of control in manufacturing. Part 2. *Soap Perfum. Cosmet.* **June** 279–283, 286.

(1983c) Microbiological aspects of control in manufacturing. Part 3. *Soap Perfum. Cosmet.* **July** 317–320, 335.

VAN ABBE, N. J., DIXON, H., HUGHES, O., and WOODROFFE, R. C. S. (1970) The hygienic manufacture and preservation of toiletries and cosmetics. *J. Soc. Cosmet. Chem.* **21** 719–800.

WARWICK, E. F. (1993) Preventing microbial contamination in manufacturing. *Cosmet. Toilet.* **108** 77–82.

WILSON, G. S. (1933) The necessity for a safe milk-supply. *Lancet* **ii** 829–832.

YABLONSKI, J. I. (1978) Microbiological aspects of sanitary cosmetic manufacturing. *Cosmet. Toilet.* **93** 37.

PART THREE

Control through preservation

5

Natural and physical preservative systems

C. MORRIS and R. LEECH

Unilever Research, Quarry Road East, Bebington, Wirral, Merseyside L63 3JW

Summary

(1)   Natural and physical preservation can be considered as a preservative system that utilizes physical preservatives, e.g. pH, natural antimicrobial agents or antimicrobial formulation components. Some of these preservative mechanisms, e.g. temperature, may impart no residual preservation on the product.

(2)   The use of natural mechanisms involves careful manipulation of the formulation and must be an integral part of the overall formulation.

(3)   Whilst there is a wide range of natural and physical preservatives, their exploitation is seldom easy because of incompatibilities with the rest of the formulation.

5.1   Introduction

During the manufacture of non-sterile pharmaceuticals, cosmetics and toiletries and during subsequent use, the product is exposed to a variety of microbial contaminants. Assuming good manufacturing practices (see chapter 4), this microbial insult from production is quite low and products are preserved to prevent any subsequent microbial growth in the formulation. Preservation can be based on the use of antimicrobial chemicals, physical factors such as pH, formulation components which show antimicrobial activity, or other systems based on biological antimicrobial mechanisms. The current trend is to move away from chemical preservation as a result of which alternative methods are being evaluated. This chapter describes the various alternatives and their applications and limitations as a means of achieving preservation.

5.2   Preservation and the physical characteristics of the formulation

Physical factors that can be considered for product preservation are hydrogen ion concentration pH, water activity a_w and redox potential Eh. It must be remembered that pharmaceuticals, cosmetics and toiletries are formulated with biodegradable compounds and as such they offer an ideal source of microbial nutrients unless they are preserved.

5.2.1 *pH*

pH, both acidity and alkalinity, can be used as a preservative mechanism but its application as the sole preservative may be limited because of the pH extremes that are required to ensure protection. The relationship between pH and hydrogen ion concentration which it represents is

$$pH = - \log_{10} [H^+]$$

where $[H^+]$ is the hydrogen ion concentration.

Micro-organisms are affected by the hydrogen ion concentration in the surrounding medium. Whilst many micro-organisms have been shown to maintain a pH homeostasis when the external pH varies (Krulwich *et al.* 1985, Zychlinsky and Matin 1983), a point is reached at which intracellular pH cannot be maintained. At this point the cell starts to lose viability. Micro-organisms have been shown to survive within the range pH 1–11 (Corlett and Brown 1980) although most optimal growth occurs for the majority of micro-organisms around neutral pH.

For non-sterile pharmaceuticals, toiletries and cosmetics the pH range for microbial survival is between pH 3 and pH 11 which unfortunately still constitutes a wide range. Within these pH limits the degree of preservation depends very much on the formulation type and the nature of the hygienic processing. The contamination risk increases as the pH approaches 7. Of course, it is often unrealistic to formulate these products at very low or very high pH and so pH control is often used in combination with other preservative systems. For example Bronopol is known to be more stable at acid pH values (pH 4–6) and as such a pH modification to the formulation to bring it within these pH values will extend the preservative life of the chemical. At pH 4 Bronopol has a 50% decomposition in over 5 years whereas at pH 8 it is only 2 months (Bryce *et al.* 1978). Another example is the use of pH depression to increase the level of undissociated acid for organic acid preservation, a method used widely in the food industry. Reducing the pH below 5 allows the formulation to be preserved with a wide range of weak organic acids, e.g. sorbic and benzoic. These acids have been shown to be potent at reducing the intracellular pH, giving an increased intracellular proton effect at a higher external pH (Salmond *et al.* 1984).

Two factors are important to remember when considering the use of pH as a means of preservation. Firstly the use of reduced pH as a preservative can encourage the survival and growth of fungi since these micro-organisms generally grow at lower pH values than bacteria (Corlett and Brown 1980). Secondly, the traditional way of measuring pH is to use a glass electrode in the test solution. This is a good way of measuring the bulk hydrogen ion concentration. However, at the microscopic level there probably exist areas of hydrogen ion deficiency and excess. It is therefore possible for some micro-organisms not to come in contact with a lowered pH environment, e.g. purely because of their size and distribution in the formulation.

5.2.2 *Water activity* a_w

Water activity of a material is defined as the ratio of the water vapour pressure of the material to that of pure water at the same temperature, i.e.

$$a_w = \frac{p}{p_0}$$

where p and p_0 are the vapour pressures of the solution and the solvent (pure water) at the same temperature.

Micro-organisms, like all living creatures, require water in an available form for survival. The availability of water can be restricted by the following methods.

(1) The amount of water in a formulation can be reduced. The obvious examples of this are powders or tablets where there is little water in the formulation.

(2) Water availability can be reduced by formulation components. Sugars and salts are two of the most frequently used groups of ingredients that are very effective at reducing water availability. Examples of formulations that can exploit this mechanism are toothpastes and pharmaceutical syrups.

The minimum a_w for growth of the major groups of micro-organisms is shown in Table 5.1. These minima assume that all other conditions are optimum for microbial survival. Within these a_w bands, further generalizations can be made. The Gram-negative bacteria that cause most microbial problems can grow at an a_w value of down to 0.90 (Christian 1980). However, for water-based cosmetics and toiletries, this Gram-negative limit has been found to be slightly higher at 0.91 (Curry 1985). The pseudomonads have a high requirement for available water and as such rarely survive below a water activity of 0.96. Below 0.91 some Gram-positive bacteria can grow, particularly *Staphylococcus aureus*, which can survive down to an a_w of 0.86. Below this water activity, yeasts and mould can grow down to an a_w of 0.6 which is the lower limit for microbial growth (Christian 1980). As bacterial spores are particularly resistant to a_w depression, they can germinate down to a water activity of 0.8 (Lewis 1969). As a consequence, varying degrees of preservation can be obtained depending on the water activity depression in the formulation. For example a formulation at an a_w of 0.95 would be preserved against pseudomonads but would still be susceptible to other bacterial and fungal contaminants.

Water activity can be depressed with a wide variety of formulation ingredients although sodium chloride, glycerol, glucose and sucrose have been most commonly tested for water binding (Curry 1985). For example a salt solution of 16.5% w/w would be required to reduce the water activity to 0.90 (Davidson *et al.* 1983) whereas 48.54% w/w glucose would be required (Taylor and Rowlinson 1955). Furthermore, it has been shown for some foods that a marked hysteresis occurs at high a_w levels. As a result, micro-organisms may grow faster in formulations with added water than those in which water is reduced, both being at the same a_w (Acott and Labuza 1975). Small differences such as this may make a large difference to the preservative capacity of an a_w, marginally preserved formulation.

Table 5.1 Limits for growth of micro-organisms

Micro-organism	a_w
Pseudomonas spp.	0.96
Other Gram-negative rods	0.90
Gram-positive rods	0.83
Moulds and yeasts	0.60

The use of water activity depression as a preservative system is well illustrated in toothpaste formulations. Toothpastes are composed basically of seven main components (Table 5.2) which vary in their importance and concentration depending on the type of paste. Whilst all the components in Table 5.2, barring the water, will lower the water activity of the final formulation, the humectant is the material which performs this function most effectively in this product type. A commonly used humectant is sorbitol which will depress the water activity as shown in Fig. 5.1. As expected, the water activity decreases as the sorbitol concentration increases.

Table 5.2   Basic toothpaste components

(1) Abrasive
(2) Binder
(3) Humectant
(4) Flavour
(5) Water
(6) Actives
(7) Detergents

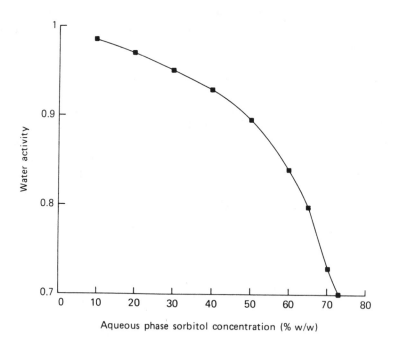

Figure 5.1   Sorbitol concentration plotted against water activity read on a Humidat-RC (Novasina)

In any formulation the amount of humectant added is of secondary importance to the aqueous phase concentration of the humectant because it is the aqueous phase concentration that dictates the water availability. This concentration can be calculated by applying the following equation:

$$\text{aqueous phase concentration of humectant} = \frac{HF \times 100}{H + W + A}$$

where H (% w/w) is the amount of humectant in the paste, F is the concentration factor of the humectant, W (% w/w) is the amount of water in the paste and A (% w/w) is the amount of any other aqueous components in the paste.

A simple toothpaste formulation, with differing levels of sorbitol, was challenged with a particularly aggressive industrial contaminant (*Enterobacter cloacae*). Fig. 5.2 shows the effect of increasing the sorbitol concentration on the survival of the micro-organism. The results depend on the time after inoculation that samples are taken but the results show a drop in viability as the aqueous phase concentration of sorbitol increases.

Thus, by altering the humectant concentration, an adequately preserved product can be obtained. However, in terms of product microbiological stability, hygienic manufacture is also an important factor when using a_w as the preservative technique. Poor-quality raw materials or poor raw material storage can lead to the presence of bacillus spores or

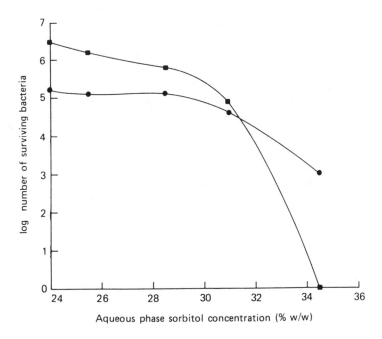

Figure 5.2 Survival of *Enterobacter cloacae* in toothpastes containing various aqueous phase concentrations of sorbitol, which had been inoculated at 2×10^8 bacteria g^{-1}. Toothpastes were sampled 1 day (•) or 7 days (■) after inoculation

osmotolerant yeast and moulds, none of which will be killed by the water activity depression used to prevent bacterial contamination. Prevention of this contamination is far more effectively handled by good hygienic processing, i.e. good-quality raw materials, storage and handling, and clean production systems, rather than relying on traditional chemical preservatives.

Other products that rely on a_w for preservation are tablets and capsules. These are discussed in detail in chapter 8.

5.2.3 Redox potential

Micro-organisms both affect and are affected by the redox potential *Eh* of the environment in which they survive and grow. Manipulation of the *Eh* not only can alter the nature of the microbial growth (Vollbrecht 1982) but also can lead to microbial death (Patel *et al.* 1984). *Eh* is a measurement of the degree of reduction or oxidation of a material and is measured in millivolts. For example *Eh* can range in different culture media from about –420 mV (anaerobic) to 300 mV (aerobic) (Brown and Emberger 1980).

Generally, micro-organisms are classed according to their general oxygen requirements, i.e. aerobic, facultative and anaerobic. Aerobes will only grow in the positive range of the redox potential. For example, *Pseudomonas fluorescens* has been reported to grow within a redox range from 500 mV to 100 mV (Oblinger and Kraft 1973). Facultative anaerobes bridge the gap between positive and negative redox potentials. The Enterobacteriaceae fall into this group, growing in a redox range from 150 mV to –600 mV (Jacob 1970). Anaerobes grow only at low redox potentials. Micro-organisms fall into these three groups depending on the terminal electron acceptors that are used for respiration. The facultative anaerobes use a variety of terminal electron acceptors and would, therefore, be the most difficult group to control using redox modification. However, the aerobic nature of most pharmaceuticals, cosmetics and toiletries tends to preclude the presence of strict anaerobes in formulations.

In the food industry, redox has been investigated for its application for preservation of meat, fruit and vegetables (Brown and Emberger 1980) and optimum redox values have been determined for particular processes, e.g. canned meats (Wirth and Leistner 1970). In any final produce the *Eh* will be determined by the product ingredients, product processing and product packaging which tends to make redox potential a difficult natural preservative system to exploit.

5.3 Formulation components and other natural preservatives

Obviously there is a large range of different ingredients in non-sterile pharmaceuticals, cosmetics and toiletries and as such it is difficult to cover the range of ingredients that may offer some preservative benefit. As a result, this section will be limited to those materials that have been widely investigated, if not exploited, for their antimicrobial activity. Essential oils and perfumes have been widely investigated as antimicrobial agents and have been reviewed by Kabara (1984). Earlier work on perfumes showed that 20% of a range of perfumery chemicals at a dilution of 1:500 inhibited at least one of the four test micro-organisms *Bacillus subtilis*, *Escherichia coli*, and two strains of *S. aureus* (Maruzzella and Bramnick 1961). Some perfumes showed inhibition down to a dilution of 1:2000 and the main groups of chemicals were ranked in order of decreasing activity, i.e. aldehydes, alcohols, miscellaneous chemicals such as Eugonol, acids, lactones, ethers, ketones, esters and acetals. These chemicals were not evaluated for antimicrobial effectiveness against pseudomonads or other Gram-negative contaminants. In more recent studies (Münzing and Schels 1972, Blakeway 1986, 1990) the major limitation with the use of essential oils for preservation was the high oil concentration required to provide an effective preservative cover. It was found that the concentrations generally required gave an olfactive limitation on the use of the oils, cosmetic and toiletry formulations having a 'medicinal' smell. This may not prevent their use in pharmaceuticals but, in these products, oil toxicity and cost will be the main factors limiting their use. Investigations of other oils has shown that they can have bactericidal and fungicidal activity. Linalool and sage oils have been shown to be fungicidal to filamentous fungi with sage oil being bactericidal and linalyl acetate bacteriostatic (Then *et al.* 1985).

Similar studies on lemon grass oil Ogunlana and Ogunlana (1985, Onawunmi 1989), velvet leaf seeds (Kremer 1986), clove and cinnamon oils (Sinha *et al.* 1992), and *Melaleuca alternifolia*, tea tree oil (Carson and Riley 1992), have demonstrated the possibilities of using oil extracts of plants as antimicrobial agents.

A review by Linton and Wright (1993) of the microbiological aspects and technological implications of the use of volatile organic compounds concluded that they have a significant role to play in inhibiting the growth of a range of micro-organisms.

A similar picture is seen with medium-chain fatty acids and esters (Kabara 1983). Again many of these substances have been tested against Gram-positive bacteria which will not necessarily reflect their activity against Gram-negative species. Some of these chemicals, e.g. monolaurin, are used specifically as preservatives, illustrating the fine dividing line between a formulation ingredient with antimicrobial properties and a chemical preservative.

Biological processes within plant and animal cells protect them against microbial attack. An understanding of these mechanisms may provide a means of naturally preserving cosmetic and toiletries against microbial contamination. Some of the protective mechanisms which have been investigated are based on enzyme systems and others on chemical interactions. Table 5.3 illustrates some of these actives. This section will concentrate on the antimicrobial nature of some of these chemicals rather than their application.

Iron binding proteins have been shown to inhibit the growth of certain bacteria (Schade and Caroline 1944). The three main iron binding proteins are transferrin from serum, lactoferrin from milk and ovotransferrin from avian egg white (Griffiths 1986). The depletion of iron from these environments to levels of 10^{-18} M is sufficient to prevent microbial survival. There are many different mechanisms, however, by which micro-organisms can survive and grow in these iron-depleted environments. Some micro-organisms have low iron requirements which reduces the effectiveness of the iron binding proteins (Reiter and Oram 1968). Others can attack and degrade the protein (Carlsson *et al.* 1984) or produce siderophores that remove iron from the protein (Perry and San Clémente 1979). This resistance has been observed in *P. aeruginosa* (Cox *et al.* 1981) which is possibly not surprising when the wide adaptability of the pseudomonads is considered.

Peroxidases are present in both plants and animals and they catalyse the reduction of hydrogen peroxide to water. Lactoperoxidase, a peroxidase found in milk and saliva, has

Table 5.3 Natural preservatives

Material	Reference	Mode of action
Lactoferrin, ovotransferrin	Griffiths (1986)	Iron binding
Lactoperoxidase	Pruitt and Tenovuo (1985)	Oxidation of SH groups
Pep 5	Sahl *et al.* (1986)	Membrane disruption
Nisin	Ruhr and Sahl (1985)	Membrane disruption
Lysozyme	Board *et al.* (1986)	Bacterial cell wall hydrolysis
Bacteriophages	Fuller *et al.* (1986)	Microbial lysis
Ovo inhibitor	Board *et al.* (1986)	Protease inhibition
Avidin	Board *et al.* (1986)	Biotin chelation

been demonstrated to produce an anti-microbial mixture when added with thiocyanate and hydrogen peroxide (Björck 1986, Perraudin 1991). Commercially available mixtures of lactoperoxidase systems have been developed for the cosmetic and toiletry business (Guthrie 1992). These mixtures have been shown to be bacteriostatic for Gram-positive bacteria and bactericidal for Gram-negative bacteria (Reiter *et al.* 1980). Processing conditions during the manufacture of cosmetics and toiletries, e.g. heating steps in creams manufacture, have been shown to denature the enzyme system; maintaining enzyme stability and activity in liquids during extended storage periods has been difficult to achieve.

There has been considerable investigation recently of the enhancement of production and mode of activity of bateriocins (Parente and Hill 1992, Vaughan *et al.* 1992, Skytta *et al.* 1992). Bacteriocins are bacterial proteins which exhibit bactericidal activity to strains and species which are usually, but not always, closely related to the producing strain of bacteria. At present the major application area for bacteriocins is in the food processing industry and there have been several reviews of the applicability of bacteriocins as natural antimicrobial agents in foods (Marugg 1991, Hillier and Davidson 1991, Wasik 1992, Morgan 1992, Eckner 1992).

Nisin, a bacteriocin produced by *Lactococcus lactis* has been widely used as a food preservative for many years. (Delves-Broughton *et al.* 1992). Nisin and Pep 5, another peptide antibiotic, have been shown to have activity against Gram-positive bacteria but little or no activity against Gram-negative bacteria and fungi. This resistance is due to lack of penetration of the peptide into the outer membrane of Gram-negative bacteria (Sahl *et al.* 1986). However, recent studies have shown that disruption of the outer membrane of Gram-negative bacteria by the use of chelating agents can make them susceptible to the bactericidal effect of nisin (Stevens *et al.* 1991).

Lysozyme works by cleaving the $\beta(1–4)$ glycosidic bonds between N-acetyl glucosamine and N-acetyl muramic acid in the microbial peptidoglycan cell wall (Board *et al.* 1986). It has been shown to be active against a range of bacteria (Chander *et al.* 1984) although antilysozyme activity does occur, most frequently in Gram-negative bacteria (Bukharin *et al.* 1984). This resistance can be due to elastase activity from the bacteria causing lysozyme inactivation (Jacquot *et al.* 1985).

5.4   Preservation and product packaging and storage

5.4.1   *Temperature*

Temperature is probably the most widely known and accepted form of physical preservation known to man. It is widely exploited for food preservation where chilled storage, pasteurization and sterilization are common methods for temperature preservation (Olson and Nottingham 1980). For pharmaceuticals, cosmetics and toiletries, temperature is not a common form of preservation because it is impracticable to store most formulations at sufficiently low temperatures to prevent microbial growth. Some pharmaceutical products have to be stored at refrigeration temperatures (4°C) or deep-freeze temperatures (–18°C) to maintain chemical stability of the product, and these storage conditions will additionally protect the product from microbial proliferation. In these cases, the storage of the product is under the control of the pharmacist with limited in-use life.

5.4.2   *Gases*

The uses of gases as product preservatives is a technique widely exploited in the food industry (Clark and Takács 1980). In addition the absence of all gases, i.e. vacuum, has also been used as a means of preservation to extend the storage life of foods (Genigeorgis 1985). Ibrahin and Sonntag (1991 and 1993) investigated the antimicrobial action of butane, carbon dioxide and dimethyl ether which are used as aerosol propellants in the cosmetic industry. They concluded that all three propellants when used individually exerted antimicrobial activity against Gram-negative and Gram-positive bacteria and against yeasts and moulds, with a significant increase in activity at 40°C compared with 20–30°C. A combination of butane/dimethyl ether with carbon dioxide could be used to protect against microbial contamination and spoilage of formulations at different pH levels and different temperatures without significant reduction in activity.

5.4.3   *Packaging*

Packaging can be classed as a preservative mechanism in that it can act as a chemical, physical and biological barrier to the outside environment. By preventing the ingress of micro-organisms, it can maintain the quality of the formulation which may otherwise be high risk in terms of microbial proliferation. This aspect is well illustrated by Sinell (1980) for food applications.

5.5   Preservation and raw materials and processing

An important factor in the attainment of acceptable microbial quality of a product is the initial microbial load. Reduction of the microbial load in a formulation can be achieved by treatment of the raw materials and/or the final product. The processes described below all have one general disadvantage in that they offer no protection to the product during any subsequent processing or during use by the consumer or patient. The maintenance of the microbial quality of the product can only be achieved by the natural antimicrobial activity of the formulation or, for processes which sterilize the product, by packaging into a sterile single-use presentation.

5.5.1 *Heat*

Although temperature may be an impracticable form of product preservation on its own, it can still play an important role in the overall quality of the final product. An examination of the heat response curve of moulds and Gram-negative bacterial species in water (Fig. 5.3) shows that quite low temperatures can effectively reduce these microbial populations. Of course, these temperatures would not be sufficient to affect microbial spores which can have a high heat tolerance (Keynan and Evenchik 1969). Heat treatment also leaves no residual preservation in the sample and so heat-treated raw materials should not be stored for extended periods without retreatment.

Heat can also be used during processing for a variety of formulations, e.g. oral liquids and skin creams. The heating step is almost always included to dissolve ingredients with a low solubility but its effect on the microbial content of the formulation should not be underrated. A switch from a hot to a cold process will increase the microbial risk to the final product (Huth 1986).

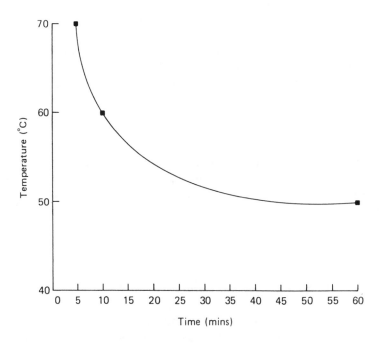

Figure 5.3   Effect of temperature on the sterilization time of water containing 10^5 cfu ml^{-1} vegetative micro-organisms

5.5.2 *Irradiation*

Radiation sources can be broadly split into two main groups depending on the emission frequency. These groups are ultraviolet irradiation and ionizing irradiation.

Ultraviolet radiation, which falls in a wavelength band between about 210 nm and 328 nm, has its maximum antimicrobial effect between 240 nm and 280 nm (Russell 1982). In a simple liquid such as water, ultraviolet radiation can effectively reduce high numbers of micro-organisms by more than four log units (Fig. 5.4) although this effect can be reduced if microbial clumping occurs. The presence of organic matter can also reduce the effectiveness of ultraviolet radiation (Morris 1972). In terms of product treatment, other factors also become important such as ultraviolet penetration through the formulation and chemical stability of product ingredients to ultraviolet exposure.

Ionizing radiations, with frequencies of 10^{18} Hz and upwards, have sufficient energy to cause microbial disruption by breaking individual molecules into oppositely charged moieties (Ingram and Roberts 1980a). These forms of radiation are β-radiation, X-radiation and γ-radiation. β-radiation is rarely used because its penetration is limited (5 meV β-rays may penetrate 2.5 cm) although the use of electron accelerators has produced some β-radiation facilities. β-irradiation can be used on some materials, e.g. polypropylene where γ-irradiation is not as acceptable. X-radiation is generally uneconomic for industrial use, which leaves γ-radiation. This is a very-high-intensity radiation which can penetrate far more than β-rays (Ingram and Roberts 1980b).

The effect of γ-radiation can be illustrated using a simple material such as a dye powder (Table 5.4). Cobalt 60 was used as the radiation source.

The cut-off point of 5,000 Gy or 0.5 Mrad agrees with the literature for the range of effectiveness of γ-radiation which is between 500 Gy and 10,000 Gy for Gram-negative bacteria and yeasts (Ingram and Roberts 1980b). A more detailed analysis of the

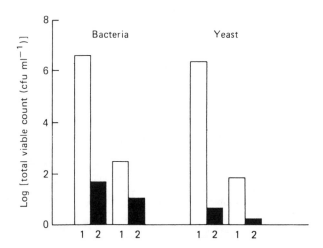

Figure 5.4 Effect of ultraviolet radiation on bacteria (*Pseudomonas cepacia* and *Enterobacter cloacae*) and a yeast (*Candida albicans*) in water at two different initial concentrations. The flow was 990 kg h^{-1} through an ultraviolet sterilizer (JAbay Ltd) with a minimum exposure of 16,000 μW cm^{-2} s^{-1} and a mean exposure of more than 30,000 μW cm^{-2} s^{-1}. The results are given prior to (1) and after (2) ultraviolet radiation treatment

surviving micro-organisms was performed using a pre-enrichment for 7 days before culturing. This technique permitted the recovery of sublethally injured micro-organisms. The enrichment was performed by weighing 1 g of powder into 9 ml of brain heart infusion broth (Oxoid) for aerobic counts or 1 g of powder into 9 ml of thioglycollate broth (Oxoid) for anaerobic counts. Similar broths were also prepared and heated to 80°C for 30 min to check for bacterial spores. These broths were incubated for 7 days at 37°C and then total viable counts were performed. The results (Table 5.5) show that the radiation was totally effective above 7500 Gy or 0.75 Mrad, illustrating the efficacy of γ-radiation. This radiation has been widely investigated for its use in product treatment particularly in the food industry, e.g. meat (Tarkowski *et al.* 1984), fish (Ito and Abu 1985) and spices (Muhamad *et al.* 1986). For pharmaceuticals, it has been suggested as a sterilization treatment for antibiotics (Tsuji *et al.* 1983) and as a treatment for drug tablets and powders (Safarov *et al.* 1985). For cosmetics, it has been investigated and used for decontaminating raw materials (Jacobs 1984).

Of course, because of the nature of this radiation, it is important to check product and ingredient stability as well as antimicrobial activity in each formulation. For example, it has been shown that some chemical preservatives are very sensitive to γ-irradiation and they can be totally lost during irradiation (McCarthy 1984). This complexity of action and activity of γ-irradiation has led to the development of models for radiation sterilization processing (Doolan *et al.* 1985, Fitch *et al.* 1985). Like temperature, there is no post-treatment preservation of the formulation when using irradiation although the radiolytic products produced during irradiation have been shown to exert some antimicrobial activity (Dickson and Maxcy 1984).

Table 5.4 Dye powder treatment with cobalt 60

Powder treatment (Gy)	Powder total viable count (cfu g^{-1})
0	4.5×10^4
2,500	4×10^2
5,000	0
7,500	0
10,000	0

Table 5.5 Recovery of micro-organisms from irradiated powder following an enrichment for 7 days in broth

Powder treatment (Gy)	Culture total viable count (cfu g^{-1})		Heated (80°C for 30 min)	
	Aerobic	Anaerobic	Aerobic	Anaerobic
0	a	a	a	a
2,500	a	a	1×10^3	0
5,000	6×10^2	5×10^1	0	0
7,500	0	0	0	0
10,000	0	0	0	0

^{a} Gross growth

80

5.5.3 Gases

The antimicrobial activity and application of gases as preservatives and more importantly as sterilants have been well reviewed (Caputo and Odlaug 1983, Christensen and Kristensen 1982, Richards *et al.* 1984). For non-sterile pharmaceuticals, cosmetics and toiletries, gases are seldom used except for the use of ethylene oxide for treating pharmaceutical raw materials and for sterilizing talcs. Its main suggested application is the treatment of formulations where no microbial proliferation occurs post-treatment, e.g. powdered drugs (Diding *et al.* 1968).

5.5.4 *Filtration*

Sterile filtration of products is a valid means of product treatment provided that the product viscosity is not too high and that downline contamination can be prevented. These constraints have limited its use for cosmetics and toiletries other than for water treatment but for pharmaceuticals it has been used widely for both water treatment and other aqueous products (Fifield and Leahy 1983). The other major disadvantage for filtration is that it gives no residual protection to the formulation. This will eliminate its effectiveness for repeat usage products.

5.5.5 *Ultra high pressure*

Hite (1899) pioneered the use of high pressure to preserve food and this technology has received renewed interest in recent years. The effect of high isobaric treatment on microorganisms isolated from foods and on the quality of food after pressure treatment has been reviewed by Hoover *et al.* (1989), Hayashi (1990), Farr (1990) and Cheftel (1991). The use of 400 MPa at room temperature for 10 minutes in the preparation of freshly squeezed orange juice inoculated with *Saccharomyces cerevisiae* showed complete elimination of the yeast and no increase in total counts during a 17-month storage period (Ogawa *et al.* 1989, 1990).

Effects of combined pressure and temperature on *Bacillus stearothermophilus* spores showed that increasing pressure from 10 to 400 MPa gave the greatest reduction in numbers of surviving spores where the temperature was greater than 60°C (Taki *et al.* 1990).

5.5.6 *High electric field pulses*

Electricity can be used to create heat or to generate oxidizing agents which can have an antimicrobial effect. High electric field pulses utilize the lethal effects of strong electric fields to inactivate microbes. The external electric field induces an electric potential over the cell membrane which, when it exceeds a critical value of approximately 1 volt, causes the formation of holes in the cell membrane by producing a repulsion between charge-carrying molecules in the membrane. For vegetative cells it is necessary to have an electric field strength of approximately 15 kV cm^{-1} to ensure the pores produced in the membrane remain apart and result in cell death. If the electric field is equal to or only slightly in excess of the critical value, it is possible for the permeability effect to be reversed and the cell membrane can be repaired resulting in survival of the cell. The critical electric field for *Escherichia coli* is 16 kV cm^{-1} (Grahl *et al.* 1990). However, the inactivation of spores would require much higher field strengths.

5.6 Conclusion

The use of natural or physical preservative systems depends on a number of constraints placed by the process and formulation being handled. For some formulations, it is impossible to incorporate natural preservation because the formulation will not remain stable. Similarly, many natural preservatives are not stable themselves and may lead to postproduction contamination owing to preservation failure. Some processing preservation methods such as heat or irradiation give no downline protection. All in all, natural preservation systems require very careful appraisal in each formulation and more importantly they require careful consideration at an early stage of product development if they are to succeed at all.

References

ACOTT, K. M., and LABUZA, T. P. (1975) Microbial growth response to water sorption preparation. *J. Food Technol.* **10** 603–611.

BLAKEWAY, J. (1986) The anti-microbial properties of essential oils. *Soap, Perfum. Cosmet.* **59** No. 4, 201, 203, 207.

(1990) Fragrances as preservatives. Cosmetic Ingredients Europe. Roure SA-55, Argenteuil, France pp. 79–90.

BJÖRCK, L. (1986) The lactoperoxidase system. In: Gould, G. W., Rhodes-Roberts, M. E., Charnley, A. K., Cooper, R. M., and Board, R. G. (eds), *Natural antimicrobial systems*. Bath University Press, Bath, pp. 297–308.

BOARD, R. G., SPARKS, N. H. C., and TRANTER, H. S. (1986). Antimicrobial defence of avian eggs. In: Charnley, A. K., Cooper, R. M., and Board, R. G. (eds), *Natural antimicrobial systems*. Bath University Press, Bath, pp. 82–96.

BROWN, M. H., and EMBERGER, O. (1980) In: Silliker, J. H. (ed.), *Oxidation–reduction potential in microbial ecology of foods*, Vol. 1. Academic Press, New York, pp. 112–125.

BRYCE, D. M., CROSHAW, B., HALL, J. E., HOLLAND, V. R. and LESSEL, B. (1978). The activity and safety of the antimicrobial agent Bronopol (2-bromo-nitropropan-1,3-diol). *J. Soc. Cosmet. Chem.* **29** 3–34.

BUKHARIN, O. V., USVYATSOV, B. Y., MALYSHKIN, A. P., and NEMTSEVA, N. V. (1984) Determination of the antilysozyme activity of micro-organisms. *Zh. Mikrobiol. Epidemiol. Immunobiol.* **2** 27–28.

CAPUTO, R. A., and ODLAUG, T. E. (1983) Sterilisation with ethylene oxide and other gases. In: Block. S. S. (ed.), *Disinfection, sterilisation and preservation*, 3rd edn. Lea and Febiger, Philadelphia, PA, pp. 47–64.

CARLSSON, J., HOFLING, J. F., and SUNDQVIST, G. K. (1984) Degradation of albumin, haemopexin, haptoglobin and transferrin by black-pigmented *Bacteroides* species. *J. Med. Microbiol.* **18** 39–46.

CARSON, C. F., and RILEY, T. V. (1992) A review: antimicrobial activity of the essential oil of *Melalenca alternifolia*. *Letters in Applied Microbiology.* **16** 49–55.

CHANDER, H., LATA, K., BATISH, V. K., and BHATIA, K. L. (1984) Antibacterial activity of lysozyme against some common food poisoning organisms. *Arch. Lebensmittelhyg.* **35** No. 4, 87–88.

CHEFTEL, J. C. (1991) Applications des hautes pressione en technologie alimentaire. *Incl. Aliment. Africae* **108** 143–153.

CHRISTENSEN, E. A., and KRISTENSEN, H. (1982) Gaseous sterilisation. In: Russell, A. D., Hugo, W. B., and Ayliffe, G. A. J. (eds), *Principles and practice of disinfection, preservation and sterilisation.* Blackwell Scientific, London, pp. 548–568.

CHRISTIAN, J. H. B. (1980) Reduced water activity. In: Silliker, J. H. (ed.), *Microbial ecology of foods,* Vol. 1. Academic Press, New York, pp. 70–91.

CLARK, D. S., and TAKÁCS, J. (1980). Gases as preservatives. In: Silliker, J. H. (ed.), *Microbial ecology of foods,* Vol. 1. Academic Press, New York, pp. 170–192.

CORLETT, D. A., and BROWN, M. H. (1980) pH and acidity. In: Silliker, J. H. (ed.), *Microbial ecology of foods,* Vol. 1. Academic Press, New York, pp. 92–111.

COX, C. D., RINEHART, K. L., MOORE, M. L., and COOKE, J. C. (1981) Pyochelin: novel structure of an iron-chelating growth promoter for *Pseudomonas aeruginosa. Proc. Natl. Acad. Sci. USA* **78** 4256–4260.

CURRY, J. (1985) Water activity and preservation. *Cosmet. Toilet.* **100** 53–55.

DAVIDSON, P. M., POST, L. S., BRANEN, A. L., and MCCURDY, A. R. (1983) Naturally occurring and miscellaneous food antimicrobials. In: Branen, A. L., and Davidson, P. M. (eds), *Antimicrobials in foods.* Marcel Dekker, New York, pp. 371–419.

DELVES-BROUGHTON, J., WILLIAMS, G. C., and WILKINSON, S. (1992) The use of the bacteriocin, nisin, as a preservative in pasteurized liquid whole egg. *Letters in Applied Microbiology.* **15** 133–136.

DICKSON, J. S., and MAXCY, R. B. (1984) Effect of radiolytic products on bacteria in a food system. *J. Food Sci.* **49** 577–580.

DIDING, N., WERGEMAN, L., and SAMULSON, G. (1968) Ethylene oxide treatment of crude drugs. *Acta Pharm. Sue.* **5** 177–182.

DOOLAND, P. T., DWYER, J., DWYER, V. M., FITCH, F. R., HALLS, N. A., and TALLENTIRE, A. (1985) Towards microbiological quality assurance in radiation sterilisation processing: a limiting case model. *J. Appl. Bacteriol.* **58** 303–306.

ECKNER, K. F. (1992) Bacteriocins and food applications. *Dairy, Food and Environmental Sanitation.* **12** No. 4, 204–209.

FARR, D. (1990) High pressure technology in the food industry. *Trends in Food Sci. Technology.* **1** 16–21.

FIFIELD, C. W., and LEAHY, T. J. (1983) Sterilisation filtration. In: Block, S. S. (ed.), *Disinfection, sterilisation and preservation.* Lea and Febiger, Philadelphia, PA, pp. 125–153.

FITCH, F. R., DOOLAN, P. T., DWYER, J., DWYER, V. M., HALLS, N. A., and TALLENTIRE, A. (1985) Towards microbiological quality assurance in radiation sterilization processing: simulation of the radiation inactivation process. *J. Appl. Bateriol.* **58** 307–313.

FULLER, R., NEWMAN, H. N., and SNOEYENBOS, G. H. (1986) Microbial competition in the mouth and gastrointestinal tract. In: Gould, G. W., Rhodes-Roberts, M. E., Charnley, A. K., Cooper, R. M., and Board, R. G. (eds), *Natural antimicrobial systems.* Bath University Press, Bath, pp. 11–28.

GENIGEORGIS, C. A. (1985) Microbial and safety implications of the use of modified atmospheres to extend the storage life of fresh meat and fish. *In. J. Food Microbiol.* **1** 237–251.

GRAHL, T., MARKL, H., SITZMANN, W., and MÜNCH, E. W. (1990) Keimabtotung mit Hilfe elektrischer Hochspannungsimpulse. Presented at the GVC-Fachausschusse Sitzung 'Lebensmittelverfahrenstechnik', Germany, 29–30 March 1990.

GRIFFITHS, E. (1986) Iron and biological defense mechanism. In: Gould, G. W., Rhodes-Roberts, M. E., Charnley, A. K., Cooper, R. M., and Board, R. G. (eds), *Natural antimicrobial systems.* Bath University Press, Bath, pp. 56–71.

GUTHRIE, W. G. (1992) A novel adaptation of a naturally occurring antimicrobial system for cosmetic protection. *Essential Oils and Cosmetics* **118** No. 9, 556, 558–560, 562.

HAYASHI, R. (1990) *Use of high pressure in food*. San-ei Publ. Co. Kyoto, Japan.

HILLIER, A. J., and DAVIDSON, B. E. (1991) Bateriocins as food preservatives. Food Research Quarterly, **51** 60–64.

HITE, B. J. (1899) The effect of pressure in the preservation of milk. *West Virginia Agric. Sta. Bull.* **58** 15.

HOOVER, D. G., METRICK, C., PAPINEAN, A. M., FORKAS, D. F., and KNORR, D. (1989) Biological effects of high hydrostatic pressure on food microorganisms. *Food Technol.* 43(3), 99–107.

HUTH, J. E. (1986) Cosmetic preservation. *Soap Cosmet. Chem. Spec.* **62** No. 3, 24–26, 78.

IBRAHIM, Y. K. E., and SONNTAG, H. G. (1991) Evaluation of the antimicrobial actions of butane, carbon dioxide and dimethylether as aerosol propellants. *Seifen-Oele-fette-wachse* **117** No. 4, 119–123.

— (1993) Effect of formulation pH and storage temperatures on the preservative efficacy of some gases used as propellants in cosmetic aerosols. *J. Appl. Bacteriol.* **74** 200–209.

INGRAM, M., and ROBERTS, T. A. (1980a) Ultraviolet irradiation. In: Silliker, J. H. (ed.), *Microbial ecology of foods*, Vol. 1. Academic Press, New York, pp. 38–45.

— (1980b) Ionising irradiation. In: Silliker, J. H. (ed.), *Microbial ecology of foods*, Vol. 1. Academic Press, New York, pp. 46–69.

ITO, H., and ABU, M. Y. (1985) Study of microflora in Malaysian dried fishes and their decontamination by gamma-irradiation. *Agric. Biol. Chem.* **49** 1047–1051.

JACOB, H.-E. (1970) Redox potential. In: Norris, J. R., and Ribbons, D. W. (eds), *Methods in microbiology*, Vol. 2. Academic Press, London, pp. 91–123.

JACOBS, G. P. (1984) Gamma-radiation decontamination of cosmetic raw materials. In: Kabara, J. J. (ed.), *Cosmetic and drug preservation*, Marcel Dekker, New York, pp. 223–233.

JACQUOT, J., TOURNIER, J. M., and PUCHELLE, E. (1985) *In vitro* evidence that human airway lysozyme is cleaved and inactivated by *Pseudomonas aeruginosa* elastase and not by human leukocyte elastase. *Infect. Immun.* **47** 555–560.

KABARA, J. J. (1983) Medium-chain fatty acids and esters. In: Branen, A. L., and Davidson, P. M. (eds), Marcel Dekker, New York, pp. 109–140.

— (1984) Aroma preservatives. In: Kabara, J. J. (ed.), *Cosmetic and drug preservation*. Marcel Dekker, New York, pp. 237–273.

KEYNAN, A., and EVENCHIK, Z. (1969) Activation. In: Gould, G. W., and Hurst, A. (eds), Academic Press, London, pp. 359–396.

KREMER, R. J. (1986) Antimicrobial activity of velvet leaf (*Abutilon theophrasti*) seeds, *Weed Sci.* **34** 617–622.

KRULWICH, T. A., AGUS, R., SCHNEIER, M., and GUFFANTI, A. A. (1985) Buffering capacity of bacilli that grow at different pH ranges. *J. Bacteriol.* **162** 768–772.

LEWIS, J. C. (1969) Dormancy. In: Gould, G. W., and Hurst, A. (eds), *The bacterial spore*. Academic Press, London, pp. 301–358.

LINTON, C. J., and WRIGHT, S. L. (1993) Volatile organic compounds: microbiological aspects and some technological implications. *J. Appl. Bacteriol.* **75** 1–12.

MCCARTHY, T. J. (1984) Formulated factors affecting the activity of preservatives. In: Kabara, J. J. (ed.), *Cosmetic and drug preservation*. Marcel Dekker, New York, pp. 359–388.

MARUGG, J. D. (1991) Bacteriocins, their role in developing natural products. *Food Biotechnology* **5** No. 3, 305–312.

MARUZZELLA, J. C., and BRAMNICK, E. (1961) The antibacterial properties of perfumery chemicals. *Soap., Perfum. Cosmet.* **34**, 743–746.

MORGAN, S. M. (1992) Potential applications for bacteriocins in the food industry. *Farm and Food* **2** No. 3, 14–15.

MORRIS, E. J. (1972). The practical use of ultraviolet radiation for disinfection purposes. *Med. Lab. Technol.* **29** 41–47.

MUHAMAD, L. J., ITO, H., WATANABE, H., and TAMURA, N. (1986) Distribution of micro-organisms in spices and their decontamination by gamma-irradiation. *Agric. Biol. Chem.* **50** 347–355.

MÜNZING, H.-P., and SCHELS, H. (1972) Potential replacement of preservatives in cosmetics by essential oils. *J. Soc. Cosmet. Chem.* **23** 841–852.

OBLINGER, J. L., and KRAFT, A., A. (1973) Oxidation–reduction potential and growth of *Salmonella* and *Pseudomonas fluorescens. J. Food Sci.* **38** 1108–1112.

OGAWA, H., FUKUHISHA, K., FUKUMOTE, H., HORI, K., and HAYASHI, R. (1989) Effect of hydrostatic pressure on sterilisation and preservation of freshly squeezed nonpasteurised citrus juice. *Nippon Nogeikagaku Kaishi* **63** 1109–1114.

OGAWA, H., FUKUHISHA, K., and FUKUMOTE, K. (1990) A method for preservation of citrus juices: a comparison of pressure and heat treatment. In: Hayashi, R. (ed.) *Use of high pressure in food*. San-ei Publ. Co. Kyoto, Japan, pp. 57–74.

OGUNLANA, G. O., and OGUNLANA, E. O. (1985) Effects of lemon grass oil on the cells and spheroplasts of *Escherichia coli* NCTC 90001. *Microbios Lett.* **28** 63–68.

OLSON, J. C., and NOTTINGHAM, P. M. (1980) Temperature. In: Silliker, J. H. (ed.), *Microbial ecology of foods*, Vol. 1. Academic Press, New York, pp. 1–37.

ONAWUNMI, G. O. (1989) Evaluation of the antifungal activity of lemon grass oil. *Int. J. Crude Drug. Res.* **27** No. 2, 121–126.

PARENTE, E., and HILL, C. (1992) A comparison of factors affecting the production of two bacteriocins from lactic acid bacteria. *J. Appl. Bacteriol.* **73** 290–298.

PATEL, G. B., ROTH, L. A., and AGNEW, B. J. (1984) Death rates of obligate anaerobes exposed to oxygen and the effect of media prereduction on cell viability. *Can. J. Microbiol.* **30** No. 2, 228–235.

PERRANDIN, J. P. (1991) Lactoperoxidase, a natural preservative. *Dairy Industries International* **56** No. 12, 21.

PERRY, R. D., and SAN CLEMENTE, C. L. (1979) Siderophore synthesis in *Klebsiella pneumoniae* and *Shigella sonnei* during iron deficiency. *J. Bacteriol.* **140** 1129–1132.

PRUITT, K. M., and TENOVUO, J. O. (1985) *The lactoperoxidase system*. Marcel Dekker, New York.

REITER, B., MARSHALL, V. M., and PHILLIPS, S. M. (1980) The antibiotic activity of the lactoperoxidase–thiocyanate–hydrogen peroxide system in the calf abomasum. *Res. Vet. Sci.* **28** 116–122.

REITER, B., and ORAM, J. O. (1968) Iron and vanadium requirements of lactic acid streptococci. *J. Dairy Res.* **35** 67–69.

RICHARDS, C., FURR, J. R., and RUSSELL, A. D. (1984) Inactivation of microorgranisms by lethal gases. In: Kabara, J. J. (ed.), *Cosmetic and drug preservation*. Marcel Dekker, New York, pp. 209–222.

RUHR, E., and SAHL, H.-G. (1985) Mode of action of the peptide antibiotic nisin and influence on the membrane potential of whole cells and on cytoplasmic and artificial membrane vesicles. *Antimicrob. Agents Chemother.* **27** 841–845.

RUSSELL, A. D. (1982) Ultraviolet radiation. In: Russell, A. D., Hugo, W. B. and Ayliffe, G. A. J. (eds), *Principles and practice of disinfection, preservation and sterilisation*. Blackwell Scientific, London, pp. 534–547.

SAFAROV, S. A., SEDOV, V. V., TYRINA, E. A., and DEIKINA, L. N. (1985) Decrease of microbial contamination of packaged drug tablets and powder by γ-radiation of ^{60}Co. *Khim.-Farm. Zh.* **3** 223–228.

SAHL, H. G., BIERBAUM, G., and KORDEL, M. (1986) Mechanism of bactericidal action of the cationic peptides nisin and Pep 5. In: Gould, G. W., Rhodes-Robert, M. E., Charnley, A. K., Cooper, R. M., and Board, R. G. (eds), *Natural antimicrobial systems*. Bath University Press, Bath, p. 321.

SALMOND, C. V., KROLL, R. G., and BOOTH, I. R. (1984) The effect of food preservatives in pH homeostasis in *Escherichia coli. J. Gen. Microbiol.* **130** 2845–2850.

SCHADE, A. L., and CAROLINE, L. (1994) Raw hen egg white and the role of iron in growth inhibition of *Shigella dysenteriae, Staphylococcus aureus, Escherichia coli* and *Saccharomyces cerevisae. Science* **100** 14–15.

SINELL, H. J. (1980) Packaging. In: Silliker, J. H. (ed.), *Microbial ecology of foods*, Vol. 1. Academic Press, New York, pp. 193–204.

SINHA, K. K., SINHA, A. K., and GAJENDRA PRASAD (1992) The effect of clove and cinnamon oils on the growth of and aflatoxin production by *Aspergillus flavus*. *Letters in Applied Micro.* **16** 114–117.

SKYTTA, E., HAIKARA, A., and TIINA MATTILA-SANDHOLM (1992) Production and characterisation of antibacterial compounds produced by *Pediococcus damnosus* and *Pediococcus pentesaceus*. *J. Appl. Bact.* **74** 134–142.

STEVENS, K. A., SHELDON, B. W., KLAPES, N. A., and KLAENHAMMER, T. R. (1991) Nisin treatment for the inactivation of Salmonella species and other Gram-negative bacteria. *Appl. and Environ. Microbiol.* **57** 3613–3615.

TAKI, Y., AWAE, T., MITSUURA, N., and TAKAGAKI, Y. (1990) Sterilisation of *Bacillus* sp. spores by hydrostatic pressure. In: Hayashi, R. (ed.), *Pressure–processed food—research and Development*. San-ei Publ. Co., Kyoto, Japan, pp. 143–155.

TARKOWSKI, J. A., STOFFER, S. C. C., BEUMER, R. R., and KAMPELMACHER, E. H. (1984) Low dose gamma irradiation of raw meat, 1, Bacteriological and sensory quality effects in artificially contaminated samples. *Int. J. Food Microbiol.* **1** 13–23.

TALYOR, J. B., and ROWLINSON, J. S. (1955) Thermodynamic properties of aqueous solutions of glucose. *Trans. Faraday Soc.* **51** 1183–1192.

THEN, M., KOSTI, E., and VERZARNE, P. G. (1985) Microbiological testing of some oils and their basic components for cosmetics. *Olaj, Szappan, Kozmet.* **34** No. 2, 41–43.

TSUJI, K., RAHN, P. D., and STEINDLER, K. A. (1983) [60]C-irradiation as an alternative method for sterilisation of penicillin G, neomycin, novobiocin and dihydrostreptomycin. *J. Pharm. Sci.* **72** 23–26.

VAUGHAN, E. E., DALY, C., and FITZGERALD, G. F. (1992) Identification and characterisation of helveticin V-1829, a bacteriocin produced by *Lactobacillus helveticus* 1829. *J. Appl. Bact.* **73** 299–308.

VOLLBRECHT, D. (1982) Restricted oxygen supply and excretion of metabolites, II, *Escherichia coli* K12, *Enterobacter aerogenes* and *Brevibacterium lactofermentum*. *Eur. J. Appl. Microbiol. Biotechnol.* **15** No. 2, 111–116.

WASIK, R. (1992) Bacteriocins, natural preservatives from bacterial compounds. *Food in Canada* **52** 29.

WIRTH, F., and LEISTNER, L. (1970) Redoxpotentiale in Fleischkonserven. *Fleischwirtschaft* **50** 491–492.

ZYCHLINSKY, E., and MATIN, A. (1983) Cytoplasmic pH homeostatis in an acidophilic bacterium, *Thiobacillus acidophilus*. *J. Bacteriol.* **156** 1352–1355.

6

The effect of container materials and multiple-phase formulation components on the activity of antimicrobial agents

G. DEMPSEY

Department of Pharmacy, King's College London, University of London, Manresa Road, London SW3 6LX

Summary

(1) Surfactant solubilized preparations and multiple-phase formulations (e.g. emulsions, creams and suspensions) pose problems of preservation against microbial attack. Similar problems may be associated with the container.

(2) Glass packaging is relatively inert and therefore unlikely to react markedly with the preservative. Evidence exists for the protective effect of glass surfaces towards absorbed microbes.

(3) Plastic containers are more widely used than glass in the packaging of pharmaceuticals. Adsorption of preservatives by plastics can be extensive, resulting in lack of protection and a serious contamination hazard.

(4) Co-solvents can enhance the solubilities of some preservatives and contribute to their antimicrobial activities. Cyclodextrins have the ability to solubilize antimicrobials but have an inhibitory effect on activity. Liposomes have found application as carrier systems for drugs. They are susceptible to microbial attack and the available concentration of antimicrobial preservative may be reduced in their presence.

(5) Preservation of aqueous suspensions of solids is influenced by the adsorption of preservative onto the solid. Subsequent antimicrobial activity depends mainly on the free aqueous phase concentration of biocide.

(6) Preservation of oil–water systems is dependent on the amount of antimicrobial remaining in the aqueous phase after the partitioning equilibrium between oil and water has been established. Stabilized oil–water products (i.e. emulsified systems) show the same effects but the emulsifying agent represents another agent for interaction with and loss of the preservative.

6.1 Introduction

Surfactant solubilized preparations and multiple-phase formulations such as emulsions, creams and suspensions can pose problems in antimicrobial preservation. Similar problems may also be associated with the container. In these situations, both the biocide and any microbial contaminants are presented with a surface and/or medium other than the formulation continuous phase. In view of this, migration of microbes to the surface with possible alterations in their susceptibility to the preservative must therefore be considered. Partitioning or adsorption of the biocide to formulation components or the container will also result in a lowering of the aqueous phase concentration, the resultant effect being loss of preservative activity.

It seems logical to postulate that where microbes and biocides concentrate at an interface, an increased lethality may result. This theory is supported by the work of Valentine (1957) and Bean *et al.* (1965). However, Allawala and Riegelman (1953) extended the Ferguson (1939) principle to aqueous antibacterial systems relating biocidal activity to thermodynamic activity. According to this principle, thermodynamic activity can be approximated to the fractional saturation of the solution. At equilibrium the thermodynamic activity of an antibacterial compound must be the same in all phases (Ferguson 1939). There is therefore no justification for assuming enhanced activity at an interface because of a surface concentration of biocide. The thermodynamic activity of biocide adsorbed at an oil–water or solid–water interface will be the same as that in the bulk aqueous phase. The increased biocidal activity as observed by Valentine (1957) and Bean *et al.* (1965) must have an alternative explanation such as a reservoir effect due to the partitioned or adsorbed preservative.

6.2 Glass and plastic packaging material

Glass has largely been replaced by plastics as a packaging material for pharmaceuticals. Thermoplastics are primarily used and include, amongst others, high- and low-density polyethylene, polypropylene, polyvinyl chloride and polyamide (Neuwald and Scheel 1969). The influence of these materials on preserved pharmaceutical products will be dependent on their interaction with contaminating micro-organisms and with any anti-microbial preservatives.

6.2.1 Glass

Glass is an inert impervious material. Interaction with preservatives must therefore be a surface adsorption phenomenon. Except where the amount of surface exposed to the product is extensive, the amount of antimicrobial lost by this mechanism will be minimal and unlikely to have any significant effect on concentration within the bulk of the product. By contrast, adsorption of bacterial contaminants by glass container surfaces producing preservative resistance is known to occur. The possibility of bacterial concentration at glass surfaces was first reported by Zobell and Anderson (1936) and Zobell (1943). Hugo *et al.* (1986) showed that the adsorption of *Pseudomonas cepacia* from aqueous suspension by glass surfaces provided protection against 0.05% w/v chlorhexidine. Increases in survival levels were correlated with exposure to larger glass surfaces. Further evidence of the protection provided by glass surfaces was shown by

Gwynn *et al.* (1981). Species of *P. aeruginosa*, *Staphylococcus aureus* and *Escherichia coli* were found to adhere to glass walls of culture vessels. This gave rise to aggregates of cells which multiplied even in the presence of bactericidal solution concentrations of β-lactam antibiotics.

The presence of contaminating organisms at the surface of glass containers even in the presence of antibacterials represents a serious potential problem. Survival and growth of these organisms may produce a considerable microbial reservoir for spoilage of the product and subsequent patient infection.

6.2.2 *Plastics*

Most data on preservative–plastic interactions refer to the packaging of sterile products such as eye-drops and contact lens solutions. With the increased use of rigid plastic containers (e.g. polypropylene and high-density polyethylene) and collapsible plastic and metal–plastic laminate tubes (Guise 1986) the problems encountered in sterile packaging apply equally to preserved non-sterile products.

6.2.3 *Plastic–preservative interaction*

Plastic containers from different manufacturers are likely to show varying preservative adsorption characteristics despite having the same plastic material as the primary component. This is the result of different processing conditions at manufacture and the presence of different additional substances such as fillers and stabilizers (Neuwald and Scheel 1969). Any attempt at quantitative comparison of preservative loss data from various researchers can therefore be meaningless and considerations must be confined to qualitative comparisons.

Extensive storage experiments by Autian (1968) on parabens solutions in plastic bottles gave an indication of likely in-use losses. Storage of aqueous solutions at 50°C for 4 months showed polyethylene containers lost 20% of methylparabens and 77% of propylparabens. Polypropylene bottles allowed 22% loss of the methyl compound and 32% of propyl.

Kakemi *et al.* (1971) studied the uptake of parabens from aqueous solutions by plastic strips. The general trend within a homologous series was increased loss with increase in ester chain length (Table 6.1). Conventional mixtures of parabens would be expected to lose the longer-chain-length component in preference to the lower homologue in proportions depending on the magnitude of the adsorption rate constants. Where significant losses occur, the residual preservative mix will have a markedly reduced bactericidal activity compared with the original solution.

The interaction of mercury-containing preservatives with plastics has been extensively studied by Fischer and Neuwald (1971). They investigated a number of polymer types from different manufacturers. The mercurial antimicrobials at bacteriostatic strengths showed varying losses over a storage period of 3 months, particularly at elevated temperatures (Table 6.2). Polypropylene and low-density polyethylene appeared less suitable as container materials than polyamide, hard polyvinyl chloride and high-density polyethylene. It is estimated that preservative losses associated with the former group of materials would be sufficient to lower the antimicrobial efficacy to an unacceptable level.

Table 6.1   The adsorption of parabens by plastic strips after 4 days at 40°C

Parabens	Amount (%) adsorbed per gram of plastic				
	High-density polyethylene	Polypropylene	Polycarbonate	Polymethacrylate	Polystyrene
Methyl	0.6	0.4	4.8	1.0	0
Ethyl	1.2	0.2	5.9	0.8	0.3
Propyl	1.5	2.4	1.4	2.9	2.8
Butyl	6.6	5.7	7.8	6.5	2.1

Initial solution strengths, 3×10^{-5} M.

Table 6.2   Storage of 0.001% phenylmercuric salts and 0.002% thiomersal solutions in different plastic containers for 3 months at 40°C

Containers	Decrease in concentration (% w/v)			
	Phenylmercuric acetate	Phenylmercuric borate	Phenylmercuric nitrate	Thiomersal
Polyamide	2.8	2.4	3.4	0.5
Hard polyvinyl chloride	0.6	3.1	2.5	0.5
Polypropylene	28.8	22.5	26.5	8.5
Low-density polyethylene (1)	45.0	25.1	36.0	10.5
Low-density polyethylene (2)	47.4	24.2	35.1	9.0
High-density polyethylene	14.8	3.1	7.6	8.5

An indication of the adsorptive capacity of nylon (i.e. polyamide) containers for preservatives other than mercurials may be deduced from the work of Guilfoyle *et al.* (1990) on adsorption by nylon membrane filters. Phenol, methyl and propyl paraben and benzalkonium chloride all showed adsorption to varying extents.

The olefin-based plastics (i.e. polypropylene and polyethylene) were shown by Autian (1968) to produce significant losses of a wider range of preservatives. These included phenol, benzalkonium chloride, phenylethyl alcohol and benzyl alcohol.

From the foregoing, it has to be concluded that the choice of plastic container for the packaging of pharmaceuticals must be an integral part of the formulation. Selection of a plastic incompatible with the antimicrobial agent present could result in a serious contamination hazard over the shelf life of the product.

6.3   Formulation components

The antimicrobial activity of biocides in formulated products is influenced by interactions with formulation components. Co-solvents and cyclodextrins in aqueous solution enhance the solubility of some antibacterials. The former have been shown to contribute to antimicrobial activity and the latter to decrease it. Suspended solids, liposomes, oil–water dispersions and surfactants in the micellar phase of a solution all represent potential systems for removal of antimicrobial components from the sphere of activity. Any contribution to activity resulting from preservatives associated with solid, oil or micelles, is likely to be minimal compared with the loss of activity due to the depleted biocide concentration in the aqueous phase.

6.3.1   Co-solvents

Co-solvents such as the alcohols, ethanol, propylene glycol and glycerol, can be used to solubilize drugs in aqueous solution (Yalkowsky and Roseman 1981). These compounds may possess antimicrobial properties and affect the activity of solubilized antimicrobials. Prickett *et al.* (1961) showed that low, non-inhibitory concentrations of propylene glycol enhanced the inhibitory effect of methyl and propyl parabens in syrup formulations against a range of vegetative bacteria, moulds and yeasts (viz. synergism). In contrast, Darwish (1992) concluded that there was no evidence that the co-solvent alcohols, cited above, caused any potentiation of parabens antimicrobial activity and furthermore that the enhanced activity of combinations could be attributed to summation of effects of the individual components. Where the co-solvent used was sorbitol, potentiation of action was evident.

6.3.2   Inclusion complexes with cyclodextrins

The cyclodextrins are cyclic oligosaccharide compounds derived by enzymic degradation of starch. β-cyclodextrin is the major product of the reaction and forms inclusion complexes with many lipophilic organic compounds (Pagington 1987, Cram 1992). Cyclodextrins have been used as adjuvants to pharmaceutical formulations to improve solubility and stability. Their potential to form inclusion complexes with antimicrobials has been shown to have an inhibitory effect on activity (van Doorne *et al.* 1988). Similarly, Loftsson *et al.* (1992) observed reduction of activity of several common preservatives by hydroxypropyl-β-cyclodextrin which agreed with the observations by Mueller *et al.* (1992) on anti-*Candida* effects of combinations of the parabens with the same cyclodextrin. In this latter case, the reduced activity was found to be dependent on the fraction of preservative lost by binding to the cyclodextrin.

6.3.3   Liposomes

Liposomes have been exploited as carrier systems for drugs, both with injectable (sterile) and topical preparations. They consist of vesicles of one or more phospholipid bilayers enclosing an aqueous layer or layers (Nässander *et al.* 1990, Brooks 1989a). Their inclusion in skin preparations by the cosmetic industry poses the problem of preservation against in use microbial attack, as with topical medicinal products. This has necessitated the inclusion of appropriate preservatives such as the parabens and diazolidinyl urea (Brooks 1989b).

Owing to the multiphase nature of aqueous dispersions of liposomes, they can represent a reservoir for removal of preservative. This was demonstrated by Komatsu *et al.* (1986) where butyl paraben was shown to equilibrate between liposomes and the continuous aqueous phase. Antimicrobial activity was proportional to the amount of free paraben. Since liposomes can attract both hydrophilic and lipophilic compounds (Nässander *et al.* 1990), a similar inactivation can be expected for other preservatives.

6.3.4 Aqueous suspensions of solids

Solid dispersions may be expected to influence preservative activity in two ways:

(1) The solid surface may adsorb any contaminating micro-organisms which may in turn affect their susceptibility to antimicrobial action (Hugo *et al.* 1986).

(2) Adsorption of the biocide will reduce the aqueous phase concentration and affect activity.

Zobell (1943) showed adsorption of bacteria by glass slides immersed in solutions of limited nutrient concentration. Concentration of nutrient at the glass surfaces stimulated growth of micro-organisms in excess of that in the bathing fluid. Enhanced metabolic activity of organisms such as *P. aeruginosa* and *E. coli* was observed in the presence of clay particles (Stotsky and Rem 1966). This was attributed in part to the buffering capacity of the clays which maintained the pH of the environment at a near optimal level. Without the clay (e.g. montmorillonite), acid metabolites quickly accumulated, lowered the pH below 5 and caused growth to cease. The adsorption of faecal organisms by kaolin and kaolin–alumina mixture (Smith 1937) of *S. aureus*, *Sarcina lutea* and *Bacillus subtilis* by kaolin (Gunnison and Marshall 1937) and *S. aureus* by kaolin (Barr 1957) has established the possibility of bacteria–clay interactions. This has been further confirmed by the work of Marshall (1968, 1969a, b) with kaolinite, montmorillonite and bentonite. Magnesium trisilicate in aqueous suspension interacted strongly with *E. coli* (Beveridge and Todd 1973), facilitating removal of the bacteria from suspension. An extensive aggregation of suspension particles with increasing bacterial cell concentration was attributed to the particle–bacteria adhesion.

During their production, inorganic powders such as kaolin, talc, magnesium trisilicate and magnesium carbonate adsorb a variety of organic compounds. Zobell (1943) and Stotsky and Rem (1966) suggested that, under these conditions, microbial contaminants concentrated at the surface could multiply at an accelerated rate. Such preparations will represent a potentially serious contamination hazard.

Adsorption of preservatives by aqueous suspensions has been investigated by several workers. Schmidt and Benke (1988) demonstrated strong adsorption of sorbic acid and benzoic acids by aluminium hydroxide. Magnesium hydroxide, magnesium trisilicate and megaldrate interacted with chlorhexidine but the magnesium trisilicate showed only slight adsorption of parabens. McCarthy (1969) found that with solids such as kaolin, magnesium trisilicate, bismuth carbonate and talc, amongst others, the amount of adsorption varied according to the solid and the preservative. Chlorhexidine diacetate (0.05%), benzalkonium chloride (0.1%) and chlorocresol (0.1%) showed greatest losses. Sorbic acid (0.1%), benzyl alcohol (2.0%) and propylene phenoxetol (1%), in contrast, were less affected. The quaternary ammonium compounds, cetylpyridinium and benzalkonium chloride were found to be adsorbed by talc and kaolin (Batuyios and Brecht 1957). The cetylpyridinium chloride was adsorbed to a greater extent by both solids. Benzoic acid was adsorbed from solution by kaolin (Clarke and Armstrong 1972) and also be sulphadimidine powder (Beveridge and Hope 1967).

Some indication of the possible influence of adsorption by a solid on the antibacterial activity of a preservative can be deduced from studies of the effect of solid adsorbents on antibiotic availability. Thus, Pinck (1962) found that montmorillonite interacted so strongly with basic antibiotics, such as streptomycin sulphate, that no biological activity could be detected by the cup plate assay technique. By contrast, the amphoteric compound chlortetracycline hydrochloride, although adsorbed, was available to show some inhibition. El-Nakeeb and Yousef (1968) in a survey of different insoluble inorganic salts on the reduction of antibacterial activity of a number of antibiotics showed varying effects depending upon the antibiotic and the suspended solid.

Preservative–suspended solid interaction and its effect on activity was demonstrated by Nakashima and Miller (1955) for benzalkonium chloride (0.1%) with 1.0% bentonite suspension. It was concluded that the benzalkonium chloride was completely inactivated. Beveridge and Hope (1967) demonstrated loss of antimicrobial efficacy of benzoic acid in the presence of sulphadimidine suspension. Bean and Dempsey (1972) showed the loss of benzalkonium chloride preservative by adsorption onto hydrocortisone suspension in eye-drops. The residual aqueous concentration of benzalkonium chloride was approximately 50% of the initial amount and was insufficient to kill an added inoculum.

More detailed studies of the relative effect of preservative in suspensions and the corresponding supernatants were carried out by Bean and Dempsey (1967, 1971). The bactericidal activity of phenol against *E. coli* in aqueous suspensions of activated carbon was found to be dependent on the residual aqueous phenol concentration after the adsorption equilibrium had been established. It was also found that aqueous suspensions of light kaolin with *m*-cresol solution showed no evidence of adsorption and activity against *E. coli* was unaffected by the kaolin. By contrast, with benzalkonium chloride, kaolin adsorbed as much as 80% of the bactericide (2.5 g light kaolin in 100 ml solution). Comparison of suspension and supernatant activity showed the suspension to have a greater bactericidal effect than the separated supernatant. This difference was more pronounced, the greater the suspension concentration, and was also seen when comparing the activities of suspensions of procaine penicillin in benzalkonium chloride solutions and separated supernatants. A similar result was obtained by Batuyios and Brecht (1957) using a suspension of talc in cetylpyridinium chloride solution. The higher activities of the solid suspensions in the presence of quaternary ammonium bactericides implies that some of the adsorbed bactericide is available to the bacteria. The observations by Bean and Dempsey (1971) emphasize that the contribution to activity by the adsorbed phase of benzalkonium adsorbed onto light kaolin is quite minimal, the predominant effect being loss of activity due to adsorption. However, the adsorbed phase contribution was proportional to the solid suspension concentration, which implied that particle–bacteria interactions were involved.

6.3.5 *Oil–water systems*

The antibacterial activity of phenolic germicides in simple oil–water mixtures has been found to be dependent on the equilibrium aqueous phase concentration (Bean *et al.* 1962, 1965). A bactericide is partitioned between the oil and water according to the partition coefficient, and the oil-to-water ratio. For any given overall concentration of biocide, the concentration in the aqueous phase is given by the following equation:

$$C_w = C \frac{\phi + 1}{K_0 \phi + 1}$$

where C_w is the concentration in the aqueous phase, C the overall concentration, K_0 the oil-to-water partition coefficient and ϕ the oil-to-water ratio. For a fixed overall concentration of preservative the effect of change in the oil-to-water ratio on aqueous phase concentration is dependent on the magnitude of the partition coefficient. An increase in the oil-to-water ratio results in an increase in aqueous phase concentration where the partition coefficient is less than unity, and a decrease where it is greater than unity (Bean *et al.* 1965). Antibacterial activity of these systems is therefore primarily dependent on the aqueous phase concentration of the preservative which is in turn determined by the partition coefficient and relative phase ratios.

A further contribution to activity is seen with these simple systems. For fixed aqueous phase concentration of the biocides phenol and chlorocresol in equilibrium with the oil phase, activity was seen to increase with increase in oil-to-water ratio (Bean *et al.* 1962). The inference from these results is that the increase in interface between the two phases is instrumental in enhancing activity. Strict adherence to thermodynamic theory as applied to antibacterial activity (Allawala and Reigelman 1953) does not allow for increased activity by an interfacial layer of bactericide. The most likely explanation of the observed phenomenon is that the oil phase preservative acts as a reservoir of material which can migrate to the aqueous environment to supplement lethality in this phase. However, the apparent contribution towards preservation from the dispersed oil phase is minimal. The most important consideration is the aqueous phase concentration of bactericide.

Simple oil–water mixtures, unless possessing self-emulsifying properties, are not stable and phase separation occurs rapidly. Conventional formulations (i.e. creams and emulsions) contain added surface-active agents which lower interfacial tension between the phases and maintain the disperse system. The presence of the surfactant introduces another source of interaction with the preservative. The non-ionic surfactants are predominantly used for formulation of stable oil–water systems in practice. In such systems the activity of any added preservative is influenced by the amount freely available in the water. This is dependent on the oil-to-water ratio, the partition coefficient and the degree of interaction with the emulgent (Bean *et al.* 1969). The following expression allows calculation of the concentration free in the water:

$$C_w = \frac{C(\phi+1)}{(K_0\phi + R)}$$

where C_w is the concentration free in the water, C the overall concentration, ϕ the oil-to-water ratio, K_0 the partition coefficient and R the ratio of total preservative to free preservative with non-ionic components in aqueous solution. With simple aqueous systems of non-ionic surfactant and preservative a plot of R against surfactant concentration C_s gives a straight line of slope S. The equation from the line is

$$R = SC_s + 1$$

The expression can be used to calculate the available preservative in different aqueous concentrations of non-ionic surfactant, assuming the slope to be independent of preservative concentration. This has been found to be so for phenol and *m*-cresol with cetomacrogol 1000 as non-ionic surfactant. The less water-soluble compounds, chlorocresol, chloroxylenol and chlorhexidine, were found to give different slopes for different total preservative concentrations (Sheikh 1971).

These aspects have also been investigated by Pisano and Kostenbauder (1959), Mitchell and Brown (1966), Bean *et al.* (1969), Kazmi and Mitchell (1978) and

Kostenbauder (1983). These workers also found that the biocidal activity of non-ionic surfactant–preservative systems was dependent on the amount of preservative free in the water. From further studies by Sheikh (1971), it was reported that, although preservative activity was related to the free preservative concentration, with cetomacrogol 1000 and bactericides such as phenol, chlorocresol and chloroxylenol, activity was always slightly greater than would be anticipated from the amount of free compound. The effect was more marked with chloroxylenol than phenol, the former compound also showing a higher degree of interaction with the non-ionic surfactant than did the latter.

Despite the apparent contribution to activity from the preservative held by the micellar phase of non-ionic surfactant, it is concluded that the paramount influence on preservative activity in aqueous non-ionic surfactant and emulsified oil systems is the amount of preservative free in the aqueous phase. Contributions to activity from preservative partitioned to the oil and surfactant are unlikely to exert a significant effect.

References

ALLAWALA, N. A., and RIEGELMAN, S. (1953) The release of antimicrobial agents from solutions of surface active agents. *J. Am. Pharm. Assoc. Sci. Edn.* **42** 267–275.

AUTIAN, J. (1968) Interrelationships of the properties and uses of plastics for parenterals. *Bull. Parenter. Drug. Assoc.* **22** 276–288.

BARR. M. (1957) The adsorption of bacteria by activated attapulgite, halloysite and kaolin. *J. Am. Pharm. Assoc. Sci. Edn.* **46** 490–492.

BATUYIOS, N. H., and BRECHT, E. A. (1957) An investigation of the incompatibilities of quaternary ammonium germicides in compressed troches. I. The adsorption of cetylpyridinium chloride and benzalkonium chloride by talc and kaolin. *J. Am. Pharm. Assoc. Sci. Edn.* **46** 524–531.

BEAN, H. S., and DEMPSEY, G. (1967) The bactericidal activity of phenol in a solid–liquid dispersion. *J. Pharm. Pharmacol.* **19** 197S–202S.

(1971) The effect of suspensions on the bactericidal activity of *m*-cresol and benzalkonium chloride. *J. Pharm. Pharmacol.* **23** 699–704.

(1972) Letter *Pharm. J.* **209** 69.

BEAN, H. S., HEMAN-ACKAH, S. M., and THOMAS, J. (1965) The activity of antibacterials in two-phase systems. *J. Soc. Cosmet. Chem.* **16** 15–30.

BEAN, H. S., KONNING, G. H., and MALCOLM, S. A. (1969) A model for the influence of emulsion formulation on the activity of phenolic preservatives. *J. Pharm. Pharmacol.* **21** 173S–181S.

BEAN, H. S., RICHARDS, J. P., and THOMAS, J. (1962) The bactericidal activity against *Escherichia coli* of phenol in oil-in-water dispersions. *Boll. Chim. Farm.* **101** 339–346.

BEVERIDGE, E. G., and HOPE, I. A. (1967) Inactivation of benzoic acid in sulphadimidine mixture for infants B.P.C. *Pharm. J.* **198** 457–458.

BEVERIDGE, E. G., and TODD, K. (1973) The interaction of *Escherichia coli* and magnesium trisilicate in aqueous suspension. *J. Pharm. Pharmacol.* **25** 741–744.

BROOKS, G. (1989a) Formulating with liposomes. *Manufacturing Chemist* **60** 36–39.

(1989b) Formulating with liposomes. *Manufacturing Chemist.* **60** 51.

CLARKE, C. D., and Armstrong, N. A. (1972) Influence of pH on the adsorption of benzoic acid by kaolin. *Pharm. J.* **209** 44–45.

CRAM, D. J. (1992) Molecular container compounds. *Nature* **356** 29–36.

DARWISH, R. M. (1992) *The effect of co-solvents on the activity of preservatives in oral aqueous dosage forms,* Ph.D. Thesis, University of London.

EL-NAKEEB, M. A., and YOUSEF, R. T. (1968) Influence of various materials on antibiotics in liquid pharmaceutical preparations. *Acta Pharm. Suec.* **5**, 1–8.

FERGUSON, J. (1939) The use of chemical potentials as indices of toxicity. *Proc. R. Soc. B,* **127** 387–404.

FISCHER, H., and NEUWALD, F. (1971) Sorption of mercury organic preservatives through plastic containers. *Pharma. Int., Engl. Edn.* **4** 11–15.

GUILFOYLE, D. E., ROOS, R., and CARITO, S. L. (1990) An evaluation of preservative adsorption onto nylon membrane filters. *J. Parenter. Sci. Technol.* **44** 314–319.

GUISE, W. (1986) Collapsible tubes. *Manuf. Chem.* **57** 44–47.

GUNNISON, J. B., and MARSHALL, M. S. (1937) Adsorption of bacteria by inert particulate reagents. *J. Bacteriol.* **33** 401–409.

GWYNN, M. N., WEBB, L T., and ROLINSON, G. N. (1981) Regrowth of *Pseudomonas aeruginosa* and other bacteria after the bactericidal action of carbenicillin and other β-lactam antibiotics. *J. Infect. Dis.* **144** 263–269.

HUGO, W. B., PALLENT, L. J., GRANT, D. J. W., DENYER, S. P., and DAVIES, A. (1986) Factors contributing to the survival of a strain of *Pseudomonas cepacia* in chlorhexidine solutions. Lett. Appl. Bacteriol. **2** 37–42.

KAKEMI, K., SEZAKI, H., ARAKAWA, E., KIMURA, K., and IKEDA, K. (1971) Interaction of parabens and other pharmaceutical adjuvants with plastic containers. *Chem. Pharm. Bull.* **19** 2523–2529.

KAZMI, S. J. A., and MITCHELL, A. G. (1978) Preservation of solubilized and emulsified systems. I. Correlation of mathematically predicted preservative availability with antimicrobial activity. *J. Pharm. Sci.* **67** 1260–1266.

KOMATSU, H., HIGAKI, J., OKAMOTO, H., MIYAGAWA, K., HASHIDA, M., and SEKAZI, H. (1996) Preservative activity and *in vivo* percutaneous penetration of butylparaben entrapped in liposomes. *Chem. Pharm. Bull.* **34** 3415–3422.

KOSTENBAUDER, H. B. (1983) Physical factors influencing the activity of antimicrobial agents. In: Block, S. S. (ed.), *Disinfection, sterilisation and preservation.* 3rd edn. Lea and Febiger, Philadelphia, PA, pp. 811–828.

LOFTSSON, T., STEFANSDOTTIR, O., FRIORIKSDOTTIR, H., and GUOMUNDSSON, O. (1992) Interactions between preservatives and 2-hydroxypropyl-β-cyclodextrin. *Drug Development and Industrial Pharmacy* **18** 1477–1484.

MCCARTHY, T. J. (1969) The influence of insoluble powders on preservatives in solution. *J. Mond. Pharm.* **12** 321–329.

MARSHALL, K. C. (1968) Interaction between colloidal montmorillonite and cells of *Rhizobium* species with different ionogenic surfaces. *Biochim. Biophys. Acta* **156** 179–186.

(1969a) Studies by microelectrophoretic and microscopic techniques of the sorption of illite and montmorillonite to *Rhizobia. J. Gen. Microbiol.* **56** 301–306.

(1969b) Orientation of clay particles sorbed on bacteria possessing different ionogenic surfaces. *Biochim. Biophys. Acta* **193** 427–474.

MITCHELL, A. G., and BROWN, K. F. (1966) The interaction of benzoic acid and chloroxylenol with cetamacrogol. *J. Pharm. Pharmacol.* **18** 115–125.

MUELLER, B. W., LEHNER, S. J., and SEYDEL, J. K. (1992) Antimicrobiological efficiency of preservatives in hydroxypropyl-beta-cyclodextrin solutions. *Pharm. Weekblad Sci. Ed.* **14** 88.

NAKASHIMA, J., and MILLER, O. (1955) Ionic incompatibilities of suspending agents. *J. Am. Pharm. Assoc., Pract. Pharm. Edn.* **16** 496, 506.

NÄSSANDER, U. K., STORM, G., Peters, P. A. M., and CROMMELIN, D. J. A. (1990) Liposomes. In: Chassin, M. and Langer, R. (eds.) *Biodegradable polymers as drug delivery systems,* Marcel Dekker, New York, pp. 261–338.

NEUWALD, F., and SCHEEL, D. (1969) The problems of packing medicines in plastic containers. *Pharma Int.* **2**, 51–54.

PAGINGTON, J. S. (1987) β-Cyclodextrin: the success of molecular inclusion. *Chemistry in Britain* **23** 455–458.

PINCK, L. A. (1962) The adsorption of proteins, enzymes and antibiotics by montmorillonite. In: Swineford, A. (ed.), *Proceedings of the Ninth National Conference on Clay and Clay Minerals.* Pergamon Press, New York, pp. 520–529.

PISANO, D. F., and KOSTENBAUDER, H. B. (1959) Interaction of preservatives with macromolecules. II. Correlation of binding data with required preservative concentrations of *p*-hydroxybenzoates in the presence of Tween 80. *J. Am. Pharm. Assoc., Sci. Edn.* **48** 310–314.

PRICKETT, P. S., MURRAY, H. L., and MERCER, N. H. (1961) Potentiation of preservatives (parabens) in pharmaceutical formulation by low concentrations of propylene gylcol. *J. Pharm. Sci.* **50** 316–320.

SCHMIDT, P. C., and BENKE, K. (1988) The adsorption and stability of preservatives in antacid suspensions. 1. Determination and influence of adsorption. *Pharm. Acta Helv.* **63** 117–127.

SHEIKH, A. W. (1971) *Studies on the influence of a surface active agent on the activity of some preservatives*, Ph.D. Thesis, University of London.

SMITH, W. (1937) A comparison between the adsorptive action of kaolin and kaolin–alumina mixture on faecal bacteria. *Lancet* **i** 438–439.

STOTSKY, G., and REM, L. T. (1966) Influence of clay minerals on micro-organisms. I. Montmorillonite and kaolinite on bacteria. *Can. J. Microbiol.* **12** 547–563.

VALENTINE, R. C. (1957) The action of chemical disinfectants on bacteria in droplets compared with that in large volumes. *J. Gen. Microbiol.* **17** 474–479.

VAN DOORNE, H., BOSCH, E. H., and LERK, C. F. (1988) Formation and antimicrobial activity of complexes of β-cyclodextrin and some antimycotic imidazole derivatives. *Pharm. Weekblad* **10** 80–85.

YALKOWSKY, S. H., and ROSEMAN, T. J. (1981) Solubilization of drugs by cosolvents. In: Yalkowsky, S. H. (ed.) *Techniques of solubilization of drugs*. Marcel Dekker, New York, pp. 91–134.

ZOBELL, C. E. (1943) The effect of solid surfaces upon bacterial activity. *J. Bacteriol.* **46** 39–56.

ZOBELL, C. E., and ANDERSON, D. Q. (1936) Observations on the multiplication of bacteria in different volumes of stored sea water and the influence of oxygen tension and solid surfaces. *Biol. Bull. (Woods Hole, Massachusetts)* **71** 324–342.

7

Formulation aspects of the preservation of hair and skin products

R. J. W. HEFFORD AND P. MATTHEWSON

Bristol-Myers Squibb, Bassington Lane, Cramlington, Northumberland, NE23 8BN

Summary

Although cosmetics have been in use for many thousands of years the science of preservation is a more recent innovation. In this chapter a holistic approach to the preservation of liquid and semi-liquid cosmetic products is suggested which should consider all aspects of the product's life from formulation through legislation to disposal. In particular product preservation is discussed from the chemical viewpoint of the formulation, its ingredients as well as the preservative system. A range of preservative types allowed in the EC is described, as is the increasing need to use combinations of materials to meet the diverse demands placed on a modern preservative system.

7.1 Introduction

The problem of preservation, against microbial spoilage, of hair and skin products has presumably been in existence for as long as the preparations have been used. Although there are no existing records of the shelf-life of asses' milk supposedly used by Cleopatra as an early form of bath salts, it is safe to assume that it was not very long and that it would have required an expiry date.

The practice of adding preservatives deliberately to products cannot have occurred before the existence of micro-organisms, as well as their unwanted effects, was recognized. However, it is quite possible that the practice of formulating products to minimize spoilage occurred through trial and error. The excellent review of Wall (1974) describes the historical progress of cosmetics in detail but does not cover the introduction of preservatives. Some good general introductions to the incorporation of preservatives into cosmetics are available (e.g. Wilkinson and Moore 1982, Schimmel and Slotsky 1974) but care must be taken as some aspects, particularly legislation, quickly become outdated. For a more recent general text see Orth (1993).

It is now generally recognized that the incorporation of a preservative system within a product is often necessary and should be a primary consideration rather than an afterthought. For efficient product development and a trouble-free produce lifetime a holistic approach must be adopted in the current climate of increased consumer awareness, legislation and cost pressures. This approach must consider the ingredients as a whole, in conjunction with consumer needs, as well as those of the manufacturer and the legislator.

In this chapter the preservation of products designed for use on two different parts of the body, the hair and the skin, is considered. These conveniently cover a range of product types exhibiting a selection of properties which offer different technical and formulation challenges.

The objective of this chapter is to consider the problem of product preservation from the chemical viewpoint of both the product and the preservative system. All the products considered have a sufficiently high water activity to be susceptible to microbial contamination. The chapter covers shampoos, conditioners, skin lotions and creams and sun protection products.

7.2 The challenge

Whenever possible the formulator should try to develop formulations which are incapable of microbial growth. However, this is difficult for the products described in this chapter. Although not all microbial contamination will cause physical harm to the consumer this can occur and, although rare, must be avoided. As discussed in chapter 2, our knowledge of the ill-effects of inadvertently using contaminated topical products is largely based upon hospital experience with pharmaceutical products; such knowledge can equally well be applied to the manufacturing and use of cosmetic products. More likely are effects such as colour change, separation, viscosity and odour changes that could cause significant damage to a brand's image and the cost of product recall (see chapter 2).

A preservative must never be used as an alternative to good manufacturing practice (GMP) and areas where contamination may be introduced must be identified and controlled. As discussed in detail in chapters 3 and 4, the principles of GMP must always be followed. Raw materials, particularly those of natural origin, must be tested for contamination before use and limits of acceptability established. In this respect water, and particularly deionized water, where micro-organisms can flourish on the exchange columns, can be a particular problem and careful attention to microbial control and monitoring should be carried out. The manufacturing facility offers a unique challenge as no two units are likely to be the same. Adequate quality assurance (QA) procedures must be in place to ensure unacceptable levels of contamination are never reached. Effective cleaning and sanitization programmes need to be validated and in place. Finally people offer an unpredictable challenge. Adequate training must be undertaken and recorded.

Once the product is made and packaged the preservative system must be able to withstand the 'normal microbial challenge'. This includes storage and use. The products discussed in this chapter may undergo a variety of challenges. Shampoos may only have transient contact with the skin whereas skin creams may be applied with the fingers. Packing should be designed to minimize the chances of contamination. The lifetime protection offered by a preservation system should be tested by a preservation efficacy test (see chapters 12 and 13).

Types of contamination vary with product type. According to the FDA definition (Blachman and Elowitz-Jeffes 1982), this becomes an issue when the organism becomes objectionable, i.e. having pathogenic potential, being present in large numbers or being deleterious to product integrity. Organisms commonly isolated from water-based products include *Klebsiella*, *Penicillium*, *Candida*, *Staphylococcus* and *Pseudomonas* spp. This last species is the most common, has very diverse metabolic capabilities, i.e. it can survive in a wide range of environments, and is often introduced through water supplies. The formulator must be aware of the types and capabilities of the organisms that may contaminate the product and use an appropriately effective antimicrobial system.

7.3 The formulation

The chemistry of liquid hair and skin products is dominated by surface and colloidal effects and it is necessary to understand a little of this science before the products can be manipulated with efficiency. An essential ingredient of all the products described here is the surfactant or surface active agent. As the name suggests, these materials collect at surfaces and are essential for cleaning and emulsion stabilization.

The behaviour of the surfactant molecule is determined by the properties of its two parts, one water and the other oil soluble. While the shape and size of these parts is important, the relative solubility of the whole molecule is described empirically by the hydrophilic lipophilic balance (HLB). A large HLB indicates high water solubility whereas a low value indicates oil solubility.

In this chapter a certain level of knowledge of surface and colloid chemistry must be assumed. However a selection of good books exists which range from introductory to master-class (Everett 1989, Hiemenez 1986).

7.3.1 *Shampoos*

Shampoos were originally simple products comprising aqueous solutions of surfactant designed to clean the hair by emulsifying oils and dispersing solid material. Subsequently a range of additional properties have been demanded by consumers and marketeers and thus provided by formulators. These include mildness, conditioning and anti-dandruff activity. The formulation of shampoos is discussed in detail elsewhere (Bendall 1989, Zviak and Vanlerberghe 1986, Shipp 1992) but the main ingredients (ignoring colours, fragrances, pH-adjusters and preservatives) are summarized in Table 7.1. For more details on ingredients, see Hunting (1985).

Shampoos are basically solutions of anionic surfactants encouraged into (at least partially) rod micellar form by cosurfactants and electrolytes in order to give a satisfactory cleaning performance and viscosity. Cationic materials can only be allowed if they can be successfully dissolved or suspended in the fundamentally anionic system.

Table 7.1 A summary of the common ingredients found in shampoos and their functions

Ingredient	Function
High HLB anionic surfactant	Cleaning/foam
Low HLB non-ionic surfactant	Foam stabilizer/thickener
Amphoteric surfactant	Mildness/foam booster
Cationic surfactant or polymer	Conditioning
Soluble/insoluble silicone	Conditioning
Insoluble stearate	Opacifier/pearlizer
Salt/glycols	Viscosity modifier

The surfactants in shampoos usually contain fatty chains with about 12 carbon atoms.

From the point of view of micro-organisms a shampoo is, therefore, a particular promising environment, with a water content of on average about 80%, a pH usually in the range 5–7 and no antagonistic ingredients.

7.3.2   Conditioners

Conditioners are designed to deposit material and are, in contrast to shampoos, basically cationic systems which are, for best effect, present as dispersions rather than solutions. A wide variety of materials may be used in conditioners (Hunting 1987, Shipp 1992). However, some usual ingredients are summarized in Table 7.2.

Conditioners are therefore usually oil/wax in water emulsions, with a cationic charge to encourage deposition which may comprise, or be suspended by, a liquid crystal phase. The viscosity control of conditioners is rather more difficult than shampoos, as it often varies with production parameters and time, and the addition of electrolytes and solvents may cause unwanted effects.

A conditioner may be a less promising environment for micro-organisms than a shampoo. Although the water content could be as high as 95%, the pH could be in the range 3–5 with the possibility of antagonistic cationic surfactants being present.

7.3.3   Skin lotions and creams

The majority of lotions and creams are emulsions of oil and water. In general a lotion is pourable whereas a cream is not. The emulsion in these products may be oil in water (o/w), water in oil (w/o) or more complex multiple systems (e.g. o/w/o). The nature of the dispersed and continuous phases is largely determined by the surfactants used to stabilize the systems, with high HLB materials used for o/w and low HLB used for w/o systems. In practice, however, more complex mixtures are often used, with mixed high and low HLB systems giving the greatest stability. The relative proportions of water and oil is secondary in determining the type of emulsion but may be important when other forces are balanced.

There are fewer references on the formulation of skin products but a number of general texts include chapters on formulation (Poucher 1984, Simmons 1990, Schmitt 1992). When consulting older general texts, it is important to ensure that such articles have not become outdated.

Table 7.2   A summary of the common ingredients found in conditioners and their functions

Ingredient	Function
Cationic surfactant/polymer	Conditioning
Low HLB non-ionic (e.g. fatty alcohol)	Cosurfactant/thickener
Wax/ester	Opacifier
Non-ionic polymer	Thickener/viscosity controller
Insoluble silicone	Enhanced conditioning

The surfactants and cosurfactants in conditioners usually contain longer fatty chains than shampoos often containing more than 16 carbon atoms.

An even wider range of materials is used in these products than in hair products, some of which are shown in Table 7.3. For products intended for cleansing or moisturization, not all ingredients will be present in any one product.

Older formulations tend to be based on fatty acid salts, often triethanolamine, whereas more recent systems utilize non-ionics. These emulsions, therefore, tend to be anionic or non-ionic in nature. Most even apparently simple products are complex emulsion systems. High-speed centrifugation of a well-known lotion has revealed four separate phases. Unusual behaviour of these systems, such as viscosity anomalies and unavailability of oil-soluble materials, may well be due to the presence of liquid crystal phases, which are known to occur in some systems (Tadros and Vincent 1983), sometimes conferring enhanced stability.

Skin-care products offer a far more varied environment to micro-organisms. They can have high water activities, with neutral pH values and ingredients as nutrients. They may offer very suitable environments as microbial growth is most commonly found at interfaces and these products have very high interfacial areas. It may be difficult for the preservative to be effective at the interface as it may not have surface activity. Conversely low water levels are possible with the water phase sometimes hidden inside an oil phase barrier. Ingredients, which at lower levels are nutrients, can at higher levels act as preservatives through osmotic effects and soaps may act in conjunction with pH values up to 10 to discourage the growth of micro-organisms.

Systems with a continuous oil phase (w/o) are generally found to be easier to preserve than o/w systems, as the exterior of the product presents an unpromising face to invading micro-organisms. However, if the product is filled hot or subject to temperature fluctuations, condensation may appear on the product offering an unprotected environment for growth.

Skin products also frequently have to withstand a high challenge due to repeated contact with fingers. The use of hand creams after bread making is an extreme example.

Table 7.3 A summary of the common ingredients found in skin lotions and creams and their functions

Ingredient	Function	Emulsion type
Salt of fatty acid (anionic)	Main emulsifier	o/w
Non-ionic surfactants	Main emulsifier	
High average HLB		o/w
Low average HLB		w/o
Fatty alcohols	Thickener	
Mineral oil	Moisturizer	
Petrolatum	Moisturizer	
Fatty esters	Emollients	
Waxes	Thickener/opacifier	
Glycols	Humectants	
Polymer (non-ionic or anionic)	Thickener/suspending	
Mg Al silicate	Suspending	
Insoluble silicone	Anti-foam/skin feel	

7.3.4 *Sun protection products*

Sun protection products are generally emulsions which contain materials capable of preventing UV radiation from reaching the skin. There are two distinct bands of wavelength of UV radiation originating from the sun which need to be considered, UV-A and UV-B. The traditional method of protection has involved the use of organic compounds capable of absorbing these two different wavelength regions. The list of materials currently allowed for use in the EC can be found in Annex VI of the Cosmetics Directive (COLIPA 1993). More recently products based on the physical blocking properties of materials such as titanium dioxide and zinc oxide have been marketed.

The environment for microbial growth in these products is very similar to that described for skin lotions and creams. However, they are regarded as more difficult to preserve (Godfrey and Seldon 1993). The potential for interaction and inactivation of the preservative with the UV absorbing materials is known to be high.

7.4 The preservative

At present there is no perfect preservative and all effective materials are a compromise of a number of often contrary properties. A brief summary of the properties that need to be considered when choosing a preservative is as follows.

(1) The spectrum of activity. This should be broad and ideally involve only one preservative at very low concentration levels. However, mixtures are becoming more necessary in order to give an acceptable performance incorporating the essential properties listed below.

(2) Safety. The materials must be safe within the context of their use and human exposure (see chapter 1). The use of risk assessment is required but it is important not to be overcritical on what might be regarded as oversensitive criteria (Godfrey 1993). Exposure in the work place in the context of COSHH must also be assessed.

(3) The question of irritation/sensitization potential is more important in some product types than others (e.g. leave-on products for sensitive skin compared to rinse-off products) and should also be considered in relation to the geographical location of use of the product (Groot and White 1992). It is likely that older materials have a worse reputation in this respect due to long-term exposure from a variety of sources.

(4) The rate of kill. This should be high to prevent microbial adaptation to the preservative system.

(5) Cost. The preservative must be cost-effective in the context of the overall product positioning, i.e. cost of goods ratio.

(6) Operational considerations. The ease of manufacture is related to safety in the workplace and is becoming increasingly important as cost and customer service pressures impact upon manufacturing. Use of a limited number of materials/suppliers, an activity usually driven by the purchasing department, within an operation may also require consideration.

(7) Environmental impact. A minor point a few years ago, this needs consideration particularly with respect to biodegradation and the effect of the product, either after its ultimate use or as waste from the factory, on water treatment plants and ecosystems.

(8) Regulatory approval. In some respects safety aspects are covered by positive lists for preservatives such as that in Annex VI of the EC Cosmetics Directive (COLIPA 1993). This is particularly important when attempting to transfer technology and to produce global formulae.

(9) Effect on the product. The preservative must not adversely affect the product properties such as odour, colour and viscosity.

(10) Functionality within the product. The preservative must not be deactivated by any of the physical properties (e.g. pH), chemical properties (e.g. ingredients such as surfactants, solids and proteins) or the packaging of the product. Interactions with packaging are discussed in more detail in chapter 6. The activity must be retained over the lifetime of the product and this has to be tested by suitable stability testing. This aspect is discussed with respect to particular preservative types later in this chapter and in more detail elsewhere (e.g. (Orth 1993, McCarthy 1984).

It is interesting to note how much this list of considerations has grown in the last 25 years and in particular how many materials are no longer considered appropriate for use in cosmetics and toiletries. If this rate of change continues, today's texts could look dated in a few years. The best way of keeping abreast of current regulations is via the membership of country trade associations, e.g. the Cosmetics, Toiletries and Perfumery Association (CTPA) in the UK and the Cosmetics, Toiletry and Fragrance Association (CTFA) in the USA, suppliers literature and reviews in the cosmetics literature (e.g. Alexander 1992, Carson 1993).

In some ways the choice of preservatives is becoming easier with the advent of positive lists. The EC positive list (COLIPA 1993) is currently divided into two parts, fully and provisionally approved. A selection of materials from this list is given in Table 7.4 and relates to materials discussed later. It is not intended to be comprehensive and some materials are omitted here for brevity.

Table 7.4 A selection of preservatives (discussed in this chapter) which are currently allowed for use in cosmetic products in the EEC as listed in and adapted from the *EC Cosmetic Directive* (COLIPA 1993)

Classification	Chemical/trivial name	Trade name
1	Benzoic acid, salts & esters	
1	Propionic acid & salts	
1	Salicyclic acid & salts	
1	Sorbic acid & salts	
4	Formaldehyde	
3	*o*-Phenylphenol & salts	
2	4-Hydroxybenzoic acid, salts & esters (parabens)	LiquaPar Oil Nipa Esters
1	Formic acid	
5	2-Bromo-2-nitropropane-1,3-diol	Bronopol Midpol 2000
3	2,4-Dichlorobenzyl alcohol	Myacide SP Midtect TF-60
3	Triclosan	Irgasan DP300

(continued)

Table 7.4 (*continued*)

Classification	Chemical/trivial name	Trade name
4	Imidazolidinyl urea	Germall 115
		Biopure 100
6	Poly(1-hexamethylenebiguanide) hydrochloride	Cosmocil CQ
		Cosmocil AF
3	2-Phenoxyethanol	Phenoxetol
4	Cis isomer of 1-(3-chloroallyl)-3,5,7-triaza-1-azoniaadamantane chloride (Quaternium-15)	Dowicil 200
4	1,3-Bis(hydroxymethyl)-5,5-dimethylimidazolidine-2,4-dione	Glydant
	(DMDM hydantoin)	Glydant XL-1000
		Nipaguard
		DMDMH
3	Benzyl alcohol	
8	1,2-Dibromo-2,4-dicyanobutane	Euxyl 400
7	5-Chloro-2-methyl-3(2H)isothiazolone & 2-methyl-3(2H)-isothiazolone (3:1)	Kathon CG
		Euxyl K100
6	Chlorhexidine salts	
6	Alkyl(c12–c22)trimethyl ammonium bromide & chloride	
8	4,4-Dimethyl-1,3-oxazolidine	Oxaban-A
4	Diazolidinyl urea	Germall II
		Germaben 2 & 2E

Provisionally listed in 1993

6	Benzethonium chloride	
6	Alkyl(c8–c18)dimethylbenzyl ammonium chloride, bromide & saccharinate	
4	Glutaraldehyde	
8	7-Ethylbicyclooxazolidine	Oxaban-E
8	3-Iodo-2-propynylbutyl carbamate	Glycacil-L & -S
		In Glydant Plus

Classification
(1) = acids, (2) = *p*-hydroxybenzoic acid esters, (3) = phenols and alcohols, (4) = formaldehyde and its donors, (5) = 2-bromo-2-nitropropan-1,3-diol, (6) = cationics, (7) = isothiazolones, (8) = miscellaneous.

Once a choice of preservative system is made, from Table 1 in the EC Cosmetics Directive, there is currently no alternative but to test whether the choice works in practice. This is done by utilizing a challenge test procedure (see chapter 12) in conjunction with a suitable stability testing regime. This should demonstrate that the preservative system retains its activity throughout the required shelf-life of the product.

In the short term a good indication of activity can be obtained using a procedure where time is accelerated by using increased temperatures. It is of course important to consider where the eventual product will be marketed and the product stored. Although tests may be undertaken using laboratory-produced material, the test should also be undertaken on product, in final packaging, manufactured under production conditions. A good monograph on this subject has been published on behalf of the International Federation of Societies of Cosmetic Scientists (IFSCC 1992). A possible test schedule is outlined in Table 7.5

Table 7.5 A possible test schedule to determine the storage characteristics of personal care products

Time (months)		Storage condition			
		4°C	20°C	37°C	50°C
0	PACX (initial)				
1		Y	Y	PA	PA
2		Y	Y	PA	PA
3		PACX	PACX	PACX	PACX
6		Y	PACX	PACX	N
12		Y	PACX	N	N
24		Y	PACX	N	N
36		Y	PACX	N	N

Where

P = Physical tests
A = Assays of actives
C = Preservative challenge test
X = Assay of preservative system
Y = Product available to test if required
N = No product available

Optional conditions

Light stability
Freeze/thaw
High humidity
Ambient

Care should be taken not to rely too heavily on accelerated test procedures as it is quite possible that processes can occur at higher temperatures that will never occur at ambient temperature. As part of a continuing QA procedure it is important that product from routine manufacture is regularly checked for preservative activity.

7.5 The preservative and the product

Table 7.4 indicates a number of general groups of preservatives and in this section the advantages and disadvantages of these materials in particular product types is discussed. A good working document which summarizes all EC allowable preservatives and their incompatibilities is available from the CTPA (1993).

7.5.1 Acids

The most important consideration in the functionality of acids is pH. At higher values the materials are dissociated and have a negative charge. The loss of activity is probably due to the presence of anionic charges on the surface of the micro-organisms resulting in repulsion. In addition, as in the case of sorbic acid activity against fungi, the material needs to be uncharged to cross the cell membrane. Acids can only function effectively under low pH conditions, e.g. as in conditioners, but unfortunately they are usually incompatible with cationics. They are commonly employed in combination with other materials.

7.5.2 p-Hydroxybenzoic acid esters

Often known as 'parabens' these materials are one of the oldest preservative types, first used in pharmaceuticals in 1924 (Orth 1993) and are still the most widely used in the USA. These materials, which are effective against Gram-positive bacteria and fungi, are often sold in combination with other preservatives active against Gram-negative bacteria. They should only be used at pH values lower than 7. Interactions occur with some ingredients such as proteins. Activity may be reduced with non-ionic surfactants, possibly due to a hydrogen bonding interaction. An antagonistic interaction occurs with sorbic acid and these materials and the combination should be avoided.

The water solubility of these esters decreases with increasing alkyl chain length whereas their functionality against fungi increases. Optimum effectiveness is usually obtained with mixtures of different chain lengths. The more effective but more oil-soluble materials may be partially inactivated by partition into the oil phase of emulsion systems. In oil continuous systems (e.g. water in oil creams) microbial growth can occur in water present on the surface of the product following condensation. Preservative present in the oil phase is, therefore, available to combat this as a proportion of the preservative will partition back into the condensate.

It seems that esters of low water solubility are quite easily solubilized within the surfactant micelle as well as partitioning into the oil phase. Although this could be regarded as a useful reservoir of extra preservative, where liquid crystal phases exist at the interfaces in emulsion systems it may be a slow process for material to move from the oil phase, if the preservative level in the water phase is depleted. This process is slow as the preservative has to pass through relatively thick layers of semi-solid material to reach its equilibrium distribution. For these materials stability and challenge testing is particular important.

The addition of metal ion chelating agents such as EDTA is thought to enhance the activity of some preservatives, particularly *p*-hydroxy benzoic esters against *Pseudomonas* (Orth (1993). This mechanism of synergy is discussed in more detail in chapter 9.

7.5.3 Phenols and alcohols

Products with a high ethanol content are self-preserving and when used below 15% ethanol may contribute towards the overall activity.

Benzyl alcohol is a naturally occurring substance which is best used above pH 5 but is deactivated by non-ionic surfactants. It is most active against bacteria and is often sold in combination with other materials where it doubles as a solvent.

Phenoxyethanol has a wide pH tolerance but is deactivated by non-ionic surfactants and polyethylene containers. It does not have a particularly broad spectrum of activity but is useful against Gram-negative bacteria. When used in emulsion systems it may drastically alter the viscosity.

Chlorophenolic derivatives such as Triclosan tend to be used for antimicrobial reasons other than preservation in products such as deodorants and in toothpastes designed to combat gum disease. This material is only sparingly water soluble and may be inactivated in structured surfactant systems and emulsions.

7.5.4 Formaldehyde and its donors

Formalin, a solution of formaldehyde in water, has a long history of successful use in cosmetics but its volatility and classification as a suspect carcinogen have resulted in a marked reduction in its use. It has a wide spectrum of activity but is incompatible with ammonium salts and proteins due to its reactive properties.

Formaldehyde donors are widely used and, although they have a quite broad spectrum, they are less active against fungi and are often sold in combination with materials such as the p-hydroxybenzoic acid esters and EDTA to increase activity. These materials tend to be compatible with other ingredients and are all quite complex cyclic compounds which hydrolyse slowly to liberate formaldehyde in solution. Generally they have a wide effective pH range and good water solubility, and are easier to handle than formalin.

Formaldehyde is known to react with the UV-A sunscreen butyl methoxydibenzoyl methane and the combination should be avoided (Godfrey and Seldon 1993).

7.5.5 2-Bromo-2-nitropropane-1,3-diol

This material, often referred to by the proprietary name Bronopol, is sometimes reported to be a formaldehyde donor but is thought not to operate via this breakdown product. It has a wide spectrum of activity, particularly against Gram-negative bacteria, is water soluble and is compatible with most ingredients. The optimum pH range for activity is 5–7. However, as a nitrite donor, this material has been suggested to assist, under appropriate conditions, in the formation of nitrosamines in the presence of secondary amines.

These limitations may be reduced by the use of a commercially available stabilized form of this material (Corbett 1993) which has less tendency to form nitrosamines and a wider pH range of activity.

7.5.6 Cationics

These quaternary nitrogen compounds suffer from inactivation by a number of other ingredients particularly anionic surfactants through a strong electrostatic interaction. They are generally more active against Gram-positive than Gram-negative bacteria. Some cationics also suffer from inactivation by absorption into plastics.

This is apparently not a problem for polyhexanide (Morpeth 1993) which operates over a wide pH range and is very water soluble. Its lower activity against fungi can be overcome by use of a combination with materials such as phenoxyethanol.

7.5.7 Isothiazolones

The commercially available product, a mixture of two materials, is the result of a chemical reaction and gives a broad spectrum preservative effective at low concentrations. The mixture is compatible with most ingredients but may be inactivated with amines and is incompatible with bisulphites and hypochlorites. This material has been the subject of controversy, due to its level of sensitization potential; there is, however, considerable available safety data to support its use. The material should not be used above pH 8.5.

The commercially available products contain varying additives and consideration must be given as to which is the most suitable for a given application. High electrolyte levels may split emulsions if addition is made in a careless fashion.

7.5.8 *Miscellaneous*

A number of more recently introduced and less easily categorized materials are worthy of mention. 1,2-Dibromo-2,4-dicyanobutane is sold in combination with phenoxyethanol and gives a broad spectrum activity and good compatibility with most ingredients. Two oxazolidine derivatives, with similar properties have broad spectrum activity and good compatibility. 3-Iodo-2-propynyl carbamate may be obtained individually or in combination with formaldehyde donors.

7.6 Conclusion

At our present state of understanding of the detailed physical chemistry of both cosmetic products and of antimicrobial action, only a rough theoretical estimate can be made of best preservative type and concentration. It is possible to avoid some gross errors. However, the empirical techniques of challenge and stability testing are still essential to confirm satisfactory activity.

The demands upon preservatives to conform to a complex ideal can only increase but it is important not to overreact to individual factors. The development and use of sensible overall risk-assessment procedures, taken in context, is very important in this respect.

References

ALEXANDER, P. (1992) Putting a stop to the rot. *Manufacturing Chemist* **63** 31–33.

BENDALL, F. (1989) *A shampoo in the making*. Hobsons Publishing in conjunction with the Society of Cosmetic Chemists.

BLACHMAN, U., and ELOWITZ-JEFFES, L. S. (1982) Microbiology of cosmetics—regulatory and quality-assurance aspects. *Cosmetic Technology* **January** 24–54.

CARSON, H. C. (1993) Preservatives. *Household and Personal Products Industries* **30** 69–76.

COLIPA, (1993) *The EC Cosmetics Directive*, April 1993 (a consolidation of directive 76/768 and its amendments up to and including the 15th adapting Commission Directive).

CORBETT, R. (1993) Stabiliser boosts biocides action. *Manufacturing Chemist* **64** 25–27.

CTPA (1993) *Guidelines for effective preservation*. The Cosmetic, Toiletry and Perfumery Association, London.

EVERETT, D. H. (1989) *Basic principles of colloid science*. Royal Society of Chemistry, London.

GODFREY, D. (1993) Maintaining the options for preservation. *Cosmetics and Toiletries Manufacturers and Suppliers* **7** 39–41.

GODFREY, D., and SELDON, C. (1993) Well preserved-sun specifics. *Soap, Perfum. Cosmet.* **66** 33–35.

GROOT, A. C. de, and WHITE, I. R. (1992) Cosmetics and skin care products. In: *Textbook of contact dermatitis*, Rycroft, R. J. G., Menne, T., Frosch, and Benezra, C. (eds). Springer-Verlag, Berlin pp. 459–475.

HIEMENEZ, P. C. (1986) *Principles of colloid and surface chemistry*, 2nd edn. Marcel Dekker, New York.

HUNTING, L. L. (1985) *Encyclopaedia of shampoo ingredients*. Micelle Press, London.

(1987) *Encyclopaedia of conditioning rinse ingredients*. Micelle Press, London.

IFSCC (1992) *The fundamentals of stability testing*, Monograph Number 2. Micelle Press, London.

MCCARTHY, T. J. (1984) Formulation factors affecting the activity of preservatives. In: Kabara, J. J. (ed.), *Cosmetic and drug preservation: principles and practice*, Marcel Dekker, New York, pp. 359–388.

MORPETH, F. (1993) Polyhexanide revisited. *Soap, Perfum. Cosmet.* **66** 37–39.

ORTH, D. S. (1993) *Handbook of cosmetic microbiology*. Marcel Dekker, New York.

PORTER, M. R. (1991) *Handbook of surfactants*. Blackie, Glasgow.

POUCHER, W. A. (1984) *Perfumes, cosmetics and soaps*, Vol. 3, revised by Howard, G. M., 8th edn, Chapman & Hall, London, pp. 310–401.

SCHIMMEL, J., and SLOTSKY, M. N. (1974) Preservation of cosmetics. In: Balsam, M. S., and Sagarin, E. (eds), *Cosmetic science and technology*, 2nd edn, Vol. 3. John Wiley, New York, pp. 391–470.

SCHMITT, W. H. (1992) Skin-care products. In: Williams, D. F., and Schmitt, W. H. (eds), *Chemistry and technology of the cosmetics and toiletries industries*. Chapman & Hall, London, pp. 99–141.

SHIPP, J. J. (1992) Hair-care products. In: Williams, D. F., and Schmitt, W. H. (eds), *Chemistry and technology of the cosmetics and toiletries industries*. Chapman & Hall, London, pp. 32–98.

SIMMONS, J. V. (1990) *The science of cosmetics*. Macmillan Education, London, pp. 115–124.

TADROS, T. F., and VINCENT, B. (1983) Emulsion stability. In: Becher, P. (ed.). *Encyclopedia of emulsion technology*, Vol. 1, pp. 129–285.

WALL, T. F. (1974) Historical development of cosmetics industry. In: Balsam, M. S., and Sagarin, E. (eds), *Cosmetic science and technology*, 2nd edn, Vol. 3, John Wiley, New York, pp. 37–161.

WILKINSON, J. B., and MOORE, R. J. (eds), (1982) *Harry's cosmetology*, Longman, Harlow.

ZVIAK, C., and VANLERBERGHE, G. (1986) Scalp and hair hygiene: shampoos. In: Zviak, C. (ed.), *The science of hair care*. Marcel Dekker, New York, pp. 46–86.

8

Preservation of solid oral dosage forms

T. C. FLATAU,[1] S. F. BLOOMFIELD[2] AND G. BUCKTON[3]

[1] Coopers & Lybrand, Management Consultancy Services, 1 Embankment Place, London WC2 6NN
[2] Chelsea Department of Pharmacy, King's College London, University of London, London SW3 6LX
[3] School of Pharmacy, University of London, London WC1N 1AX

Summary

(1) Tablets and dry raw materials may contain high numbers of viable micro-organisms. The risk of infection from contaminated solid dosage forms is discussed and pharmacopoeial limits for oral solid formulations and raw materials are reviewed.

(2) The availability of water to micro-organisms is explained in terms of water activity (a_w) and equilibrium relative humidity (ERH). Product spoilage is more likely to be due to moulds than to yeasts or bacteria, since moulds have lower water requirements for growth.

(3) Solid dosage forms are essentially self-preserved so long as they are kept dry. The availability of environmental moisture to microbial contaminants depends on the nature of the interaction between sorbed water and the formulation components.

(4) For product manufacture involving a granulation step, a high degree of kill can be achieved by granule drying even at fairly low temperatures. Tablet compression has been found to kill contaminating micro-organisms.

(5) Preservatives for tablets have been studied, but are unlikely to be justifiable to licensing authorities, since effective preservation of solid oral dosage can be achieved through the use of high-quality raw materials, good manufacturing practice (GMP) and formulation, packaging and storage conditions which discourage water uptake.

8.1 Introduction

Historically, the dry nature of solid oral dosage forms has led to the assumption that such formulations are not at risk from microbial spoilage. However, a number of reported infection outbreaks associated with tablets and capsules indicate that this is not necessarily the case. Even vegetative pathogenic bacteria can survive in apparently dry

113

Microbial quality assurance in cosmetics, toiletries and pharmaceuticals

pharmaceutical products and some raw materials are comparatively more likely to sustain contamination.

8.2 Contamination of pharmaceutical solid dosage forms

8.2.1 Incidences of contamination

The microbiological quality of non-sterile pharmaceuticals was first given serious attention following a report by Kallings *et al.* (1966a) to the Swedish National Board of Health, which described contaminated tablets containing numbers of coliforms 'up to millions per gram'. Publication of the report was closely followed by an outbreak of *Salmonella muenchen* infections in Sweden. Following an epidemiological investigation into the outbreak, the infection was traced to defatted thyroid powder containing over 3×10^7 bacteria per gram, mostly faecal flora (Kallings *et al.* 1966b).

In the USA the Center for Disease Control (1966) documented an outbreak of salmonellosis involving one infant death. The infections were traced to carmine powder (which was not in fact intended for oral consumption) contaminated with *Salmonella cubana*. The powder had been hand-filled into gelatin capsules for use as a faecal dye. This provided one of the early reports in the literature of contaminated gelatin capsules. Such serious consequences of contaminated products led to further surveys into the microbial content of solid oral dosage forms. Extensive surveys of tablet contamination in Denmark were conducted by Fischer *et al.* 1968) and Fuglsang-Smidt and Ulrich (1968). A total of 696 batches of 237 tablet types were examined, using non-specific viable counting techniques. Only 53 batches contained more than 100 bacteria per tablet, although counts up to 10^5 organisms/tablet were recorded for three batches of thyroid tablets. Qualitative methods failed to show the presence of Salmonellae or *Escherichia coli*.

In surveys of antibiotic drugs, White *et al.* (1968) reported gross contamination of an antifungal powder by an aerobic spore-forming organism (*Bacillus* species). Bowman *et al.* (1971) examined market samples of antibiotic preparations. Of the tablets and capsules tested, none contained more than 50 organisms per gram and no specific pathogens were identified.

Jain and Chauhan (1978) reported appreciable levels of contamination in batches of antacid tablets in India, with a mean total viable count of 4.36×10^3 per gram. Contamination levels were found to be higher for batches made by small-scale manufacturers than large-scale manufacturers, the consequences of inferior production facilities and control. A similar study of market samples of tablets in India showed almost all samples to contain fungal contamination, amongst which *Candida albicans*, *Aspergillus niger* and *Aspergillus flavus* were identified. Bacterial contamination was found to be up to 1.16×10^4 per gram (Khante *et al.* 1979). Somerville (1981) surveyed non-sterile pharmaceuticals manufactured in Rhodesia and found the tablets examined to be microbiologically satisfactory, since potentially pathogenic organisms were not identified. Out of 174 samples, 35 showed growth, but over half of the contaminated samples contained less than 100 organisms per gram. However, in a later survey, pathogenic bacteria were found in many of the tablets sampled from an Indian hospital pharmacy (Nandapurkar *et al.* 1985). Garcia-Arribas *et al.* (1986) carried out an extensive study of contamination of oral solid dosage forms. Of 240 samples, over 30% were contaminated with aerobic bacteria, mostly *Bacillus* species, from which 18 species were identified.

Since solid oral dosage forms are dry, they offer an inhospitable environment for microbial growth; it is unlikely that organisms introduced during processing will flourish. The major risk is rather from the indigenous population of contaminants in raw materials, both active drug substances and excipients which may be common to many formulations. Schiller *et al.* (1968) and Pedersen and Ulrich (1968) surveyed microbial contamination in pharmaceutical raw materials and found natural products, especially those containing extracts of natural origin, to have the highest counts. In these surveys, one batch of digitalis leaf contained 10^7 viable organisms per gram, and low levels of *E. coli* (30–100 per gram) were found in lactose and talc, prednisone and prednisolone. Earlier, Kallings *et al.* (1966b) had shown that in tests on a large number of raw materials, starch had been the most commonly contaminated, with potato or wheat starch more at risk from coliforms than maize starch. In similar surveys, Kruger (1973), and later Garcia-Arribas *et al.* (1983) found that the more contaminated products were of natural origin. Wozniak (1971) had found that oral medicines derived from plants contained up to 10^7 organisms per gram and 75% of samples contained *Bacillus cereus*. Coagulase-positive *Staphylococcus aureus* was frequently isolated. Table 8.1 summarizes some of the reports of contaminated products.

8.2.2 The hazard to the user

Most of the bacterial and fungal species isolated from dry pharmaceutical products do not present a direct hazard of infection to the patient. However, the presence of known pathogens and bowel flora in dry finished products and their raw materials has been shown in a number of the surveys referenced above (section 8.2.1). It is perhaps surprising that more cases of infection from solid oral dosage forms have not arisen.

As discussed in chapter 2, the risk of infection associated with ingesting a contaminated product depends on a number of factors, including the virulence and pathogenicity of the organisms concerned, the condition of the patient and the total number of organisms present. The doses of orally administered pathogenic organisms required to cause disease have been studied. McCullough and Eisele (1951) administered a variety of *Salmonella* species and strains to human volunteers, and showed that depending on the strain, between 1.6×10^5 and 2.5×10^7 organisms could be given as a dose without

Table 8.1 Some results showing the level of microbial contamination of tablets

Reference	Number of batches examined	Number of batches with the following numbers of viable organisms per gram					
		<10	10–10^2	10^2–10^3	10^3–10^4	10^4–10^5	>10^5
Kallings *et al.* (1966b)	157	119	4	6	8	4	16
Fuglsang-Smidt and Ulrich (1968)	360	247	89	18	5	1	—
Somerville (1981)	174	146	15	8	3	2	—

causing illness. However, an outbreak of salmonellosis was traced by Coyle *et al.* (1988) to home-made ice cream containing *S. enteridis* $2.1–5.0 \times 10^4$ per gram, and reported outbreaks involving chocolate and Cheddar cheese suggested that the infective dose was as little as 50–100, and less than 10 organisms respectively (Gill *et al.* 1983; Greenwood and Hooper, 1983; D'Aoust, 1985).

Schneller (1978) reviewed reports of work involving the administration of pathogenic strains to healthy volunteers. When *E. coli* was fed to healthy adult males (Ferguson and June, 1952), 7×10^6 organisms gave only slight illness in 7 out of 10 subjects, with no definite diarrhoea. Similarly, *Pseudomonas aeruginosa* was administered at doses up to 2×10^8 with recovery of the organism in the stools but no signs of clinical illness (Buck and Cooke, 1969).

In a review of microbial contamination in oral dosage forms, Wallhausser (1977) drew attention to mycotoxins and bacterial toxins, which are capable of exerting poisonous effects at extremely low concentrations. Mycotoxins, and particularly the aflatoxins because of the universal occurrence of *A. flavus*, probably represent the greatest hazard. According to Wallhausser, bacterial toxins are equally noxious but largely overlooked. *Bacillus* species are the most commonly isolated microbial contaminants from pharmaceutical products (Garcia-Arribas *et al.* 1986). Although generally considered to be non-pathogenic, *Bacillus* species have been reported to mediate serious infections, associated with exotoxin production (Farrar 1963, Turnbull *et al.* 1979). Furthermore, the majority of strains isolated from pharmaceuticals are resistant to many antimicrobial agents (Garcia-Arribas *et al.* 1988). These organisms are also common laboratory contaminants, so that *Bacillus* isolations may be incorrectly attributed to poor experimental technique. Consequently, it is possible that illness with *Bacillus* involvement may be more common than the literature suggests.

8.2.3   Standards for the microbial content of pharmaceutical products

The question of safety of contaminated pharmaceuticals, as well as the effect of micro-organisms upon the quality and efficacy of medicinal products, has resulted in the introduction of limits for the microbial quality of tablets and other non-sterile pharmaceuticals. The problems of specifying suitable limits is discussed in detail in chapter 15. Table 8.2 summarizes compendial limits applied to solid pharmaceuticals. In brief, the standards for microbial quality adopted by licensing authorities and manufacturers have been based on two requirements: the complete absence of specified micro-organisms, or a restriction on total numbers of permitted micro-organisms. It is recognized that this latter standard may be difficult to impose, since the size of a population of contaminants will vary according to the time or stage of processing at which the product is tested, and the method of testing. However, this approach has been adopted by the *European Pharmacopoeia* (Anon. 1980), amongst others. Limits on total numbers are also used extensively by manufacturers for 'in-house' quality monitoring of raw materials and finished products.

In general, official requirements for the complete absence of specified bacteria apply only to specific 'at-risk' products, e.g. the *British Pharmacopoeia* (Anon. 1993a) considers digitalis and gelatin to be among those products which are required to be free from *E. coli*, *Salmonella* species and pseudomonads. The *United States Pharmacopeia* XXIII (Anon. 1995) states: 'The nature and frequency of testing vary according to product. The significance of micro-organisms in non-sterile pharmaceutical products should be

Table 8.2 Suggested microbial limits for solid pharmaceuticals

Reference	Material or product	Requirements
British Pharmacopoeia (1993a)	Specifies 'at-risk' materials	Absence of *Escherichia coli*, *Salmonella* spp. and *Pseudomonas* spp.
United States Pharmacopoeia (1985)	Nature and frequency of testing vary according to product, e.g. natural products	Absence of *Salmonella* spp.
European Pharmacopoeia (1995)	Oral solid	$\not> 10^3$ bacteria g^{-1}; $\not> 10^2$ fungi g^{-1}; Absence of *Escherichia coli*
Fédération Internationale de Pharmaceutique (1975)	Oral solid	10^3–10^4 aerobic bacteria g^{-1}; $<10^2$ moulds g^{-1} or yeasts g^{-1}

evaluated in terms of the use of the product, the nature of the product and the potential hazard to the user. It is suggested that certain categories of products should be tested routinely for total microbial count and for specified microbial contamination, e.g. natural plant, animal and some mineral products for *Salmonella* species.'

8.3 Water activity and microbial survival in solid dosage forms

8.3.1 The basic concepts of water activity

Micro-organisms will only grow in aqueous solutions where nutrient and other conditions are favourable. The growth solution may be extremely dilute, for example distilled water (Bigger and Nelson 1941), or potentially very concentrated, as in the case of solid and semi-solid substrates. Growth in solid dosage forms may be considered as a special case of growth in a concentrated solution. In studying microbial survival and potential growth, it is important to appreciate the properties of aqueous solutions as they affect the availability of water to micro-organisms.

It has long been shown that foods can be preserved by reducing their water content, and that spoilage proceeds more rapidly under humid conditions. However, it has been noted that spoilage can occur in some systems of low water content, but need not occur in some systems containing relatively more moisture, for example sucrose containing 4% water (dry basis) was liable to spoilage, whereas potato starch containing 24% moisture (dry basis) was perfectly stable (Van den Berg and Bruin 1981). Clearly, it is not the absolute water content, but some other factor relating to the nature of constituent water that determines the probability of spoilage.

Scott (1957) derived the term 'water activity' to explain how the status of a solution can affect the availability of constituent water to micro-organisms. If a solid oral dosage form is regarded as a potentially saturated solution dispersed on a solid substrate, then the status of water in this solution determines the potential to support life. In an aqueous solution, the orientation of water molecules around the solute results in a decrease in entropy, and a consequent increase in intermolecular forces. This increase in the inter-molecular forces is reflected and can be monitored by a lowering of the vapour pressure above the solution. For an 'ideal' solute, the extent of vapour pressure reduction is given by Raoult's law:

$$p/p_o = N_2/(N_1 + N_2) \qquad (8.1)$$

Where p and p_o are the vapour pressures of the solution and solvent respectively at the same temperature, and N_1 and N_2 are the number of moles of solute and solvent, it can be seen that the vapour pressure above the solution (dosage form) compared with that above the pure solvent (water) is equal to the mole fraction of the solvent. The value obtained from equation (8.1) expressed as a percentage will be the vapour pressure observed above the solution once an equilibrium has been achieved, i.e. the equilibrium relative humidity (ERH). The following definitions may be of value.

(1) Humidity is the mass of vapour associated with a unit mass of gas.
(2) Percentage humidity is the ratio of humidity to the humidity of the saturated gas at the same temperature, expressed as a percentage.
(3) Relative humidity (RH) is the ratio of the partial pressure of the vapour in the gas to the partial pressure when the gas is saturated, expressed as a percentage.

The value obtained from equation (8.1) expressed directly is termed the water activity a_w, which is 100 times lower than the relative humidity. Water activity is strictly a solution property, and relative humidity is a vapour property. A molar aqueous 'ideal' solution will have an a_w of 0.9823, and is capable of producing an ERH of 98.23%. Thus ERH is an indirect measure of a_w.

8.3.2 Water interactions with solids

The quantity and nature of water in a solid dosage form will depend on the environment in which it is stored and the physical properties of the product. Tablets usually contain a small amount of water, which acts as an aid to compaction. On storage, this water may be lost to a dry atmosphere or supplemented from a humid atmosphere, until the solid reaches its characteristic equilibrium moisture content (EMC) at the temperature and ERH of the vapour environment. The EMC of a solid formulation will be the weighted mean of the EMC values of the various components. The term 'hygroscopicity' relates to both the EMC of a solid, and the rate at which it is achieved. At the EMC extremes for example, amylobarbitone, a hydrophobic powder, when stored at 100% ERH will adsorb less than 1% w/w water and apparently result in no dissolution (Buckton *et al.* 1986), whilst a deliquescent powder would rapidly form a saturated solution which would fit more easily with the thesis of Scott (1957).

However, all adsorbed water is not available for use by contaminating micro-organisms. Water has a tendency to interact with solids in a variety of ways, including adsorption onto the surface to form a monolayer, further adsorption to form a multilayer, capillary condensation into pores, absorption into a lattice to form a hydrate (which may or may not be stoichiometric), and absorption into amorphous regions of a solid.

8.3.2.1 Water adsorption

Adsorbed water vapour is bonded to the substrate surface by hydrogen bonds, t
of which is determined by the alignment of the water molecules. It is clear that
adsorption of water occurs on water-soluble materials without the solid dissol
the water that is adsorbed cannot be behaving as liquid water. There has be
discussion about the nature of adsorbed water. The Brunauer, Emmett and Teller ᴄquation
(Brunauer *et al.* 1938) distinguishes between bound water, i.e. that which binds to the
solid, and all subsequent sorbed water, which should be regarded as condensed water.
Another model is described by the GAB equation (developed independently by
Guggenheim (1966), Anderson (1964) and de Boer (1968). The GAB model treats water
adsorption as a three-stage process, where there is monolayer adsorption, condensed
water, and an intermediate state between the two extremes. In general, the GAB model
provides a better fit to most experimental data.

Many solids of pharmaceutical interest take up more water than can be explained by
simple adsorption (given that multilayer adsorption does not progress beyond 3–5
molecules thick). It follows that if the water is not taken up by adsorption, then it is either
condensing into pores, or if no such pores exist, it must be absorbing to the solid. It is
well known that water absorbs into amorphous materials.

8.3.2.2 Capillary condensation

Capillary condensation will result in significantly greater amounts of water adsorbing to
microporous materials than would occur for smooth surfaces. Adsorption occurs in
micropores due to the formation of a meniscus at the entry to the pore, which will have a
radius of curvature, such that there will be a pressure difference that will change the local
a_w. This effect is described by the Kelvin equation:

$$\ln p/p_o = 2\gamma M/(RT\rho r) \tag{8.2}$$

where γ is the interfacial tension, M is the molecular weight, ρ is the density of water, R
is the gas constant, T the absolute temperature and r the radius of curvature of the liquid
entering the pore. It can be calculated that condensation will occur in pores of 100 nm
when the relative humidity is 99%, but at only 50% for pores of 1.5 nm. It follows that
condensation will occur in pores at much lower relative humidities than are needed over
the surface of the material.

The adsorption of water into pores is especially vulnerable to hysteresis, since the
pressure difference across the meniscus, which facilitates adsorption, will often work to
hinder the desorption of water. This can have a profound effect on the water content of a
material.

8.3.2.3 Water absorption into amorphous regions

Given that many non-porous surfaces also show water sorption in excess of that which
can be accommodated in a few layers on the surface, it can be deduced that absorption is
relatively common. Many pharmaceutical materials are regarded as crystals, but in fact
the surfaces can be made partially amorphous by processing. Sorption into amorphous
regions can account for the excess water uptake, and explains the extensive hysteresis
that is often observed.

The effects of water sorption into partially amorphous materials, have been considered by Ahlneck and Zografi (1990). These authors discussed the consequences of the bulk of water associated with solids being present in the small amorphous regions. For example, a solid which is 0.5% amorphous, and which has been shown to sorb 0.5% w/w of water, would normally be considered to have an insignificant water content at this level. However, if all the sorbed water is contained in the amorphous material, then the local environment would be a 1:1 ratio of water to solid. Ahlneck and Zografi (1990) have shown that only 0.1% moisture is needed in a sucrose sample which is 1% amorphous, in order to plasticize the amorphous material below its glass transition temperature, and thus totally change the material properties. Such changes in local physical properties make these regions more susceptible to the onset of chemical degradation reactions, and also to physical transitions. It is equally true that local high a_w conditions in amorphous regions may be far more likely to support microbial growth.

8.3.2.4 Crystal hydrates

Stoichiometric crystal hydrates have water chemically bonded into the crystal structure which will not be available for use by micro-organisms. However, different hydrates have different conditions of temperature and humidity at which they will give up their water of crystallization. A knowledge of changes in the crystal form (hydrate/anhydrous) will be important, since the released water of crystallization will be free for use in product spoilage. Some materials give up water freely at low temperatures, causing the crystal to collapse. Non-stoichiometric hydrates also exist, in which water is held in a crystal lattice, without altering the molecular packing of the solid. These solids are thought to contain channels into which water can fit, and from which it can also be removed. Water held in this form is neither adsorbed, nor chemically bonded into the lattice.

8.3.2.5 Investigation of water interactions

There are many ways by which materials can be studied with respect to their physical form and the corresponding interaction with water. Sorption processes can be followed gravimetrically, and presented as isothermal plots of moisture content versus pressure or a pressure function, such as ERH or a_w. Commonly, materials of pharmaceutical interest follow an isotherm with two inflection points. The first (lower humidity) inflection is thought to correspond to the point at which a monolayer of adsorbed molecules has formed on the solid. The second (higher humidity) inflection is regarded as the point at which the sorbed water has plasticized the solid such that the glass transition temperature of the material has been reduced to the temperature of the experiment. It follows that water sorbed above the second transition point will be far more readily available than sorbed water below this point.

A variety of thermal methods can be used to assess the state of water associated with solids. Differential scanning calorimetry can be used to ascertain whether the water is freezable (i.e. free) or bound in a more restricted form. This has been applied to water uptake by hydrocolloids by Bhaskar *et al.* (1993). Isothermal microcalorimetry can be used to study water sorption and desorption (e.g. Sheridan *et al.* 1993). This methodology has been applied to moisture adsorption for pharmaceutical starches and microcrystalline celluloses (Blair *et al.* 1990) to show how water binds to anhydroglucose units in the amorphous regions of the solid. Desorption can be followed by isothermal microcalorimetry, and by techniques such as thermogravimetric analysis. The thermodynamics and energetics of desorption can be quantified and the extent of physical and chemical bonding determined. With such approaches, the experiments are designed to investigate both the extent of water interaction and the nature of the bonding. Tightly bound water will not behave as bulk liquid, whilst loosely bound water will have a far greater potential for interactions with other phases.

Assessment of the ERH above a solid product can be regarded as a measure of the free (loosely bound) water, which may be available to micro-organisms. Strongly bound water cannot be regarded as being in solution, and consequently is not involved in the expression for available water (equation (8.1)).

8.3.3 The effect of water activity on survival and/or growth of microbial contaminants

8.3.3.1 Moulds

Scott's 1957 review summarized early findings concerning mould germination, growth and a_w. At each a_w there was a characteristic growth rate constant, which decreased as the a_w was reduced (Tomkins 1929). For spore germination, it was reported that a reduction in a_w caused a reduction in germination rate, and that each fungus had an optimum temperature at which the rate of germination was greatest. Of 20 cultures of *Aspergillus* species, three germinated at $0.70a_w$, seven at $0.80a_w$ and ten at $0.85a_w$ (Galloway 1935). Heintzeler (1939) reported that latent periods before germination could be as long as 70 days at lower water activities, with spores from younger cultures germinating best. For different mould spores the minimum a_w required for germination varied from 0.70 to 0.98 (Bonner 1948). For optimum germination, Ayerst (1969) reported that temperatures may vary from 23°C to 40°C, and the optimum a_w may vary from 0.93 to greater than 0.98.

Spoilage problems associated with solid pharmaceuticals are most likely to involve xerophilic fungi which, according to Pitt (1975), are those species 'capable of growth, under at least one set of environmental conditions, at a water activity below 0.85'. The xerophiles listed in Pitt's review include many *Aspergillus* species and *Penicillium* species, together with *Xeromyces bisporus* which can justifiably claim the lowest a_w at which fungal growth has been reported, at just $0.61a_w$. Studies of survival of fungi on pharmaceutical tablets at lower a_w (Fassihi and Parker 1977) suggest that storage at $0.36a_w$ will result in mortality after about 4 weeks, but storage at $0.44a_w$ will allow viable spores to survive for at least 7 weeks (the duration of the experiment). Thus under normal temperate storage conditions, defined as 21°C/45% RH (Anon. 1993b), fungal spores may remain viable in pharmaceutical solids. However, it appears that sustained high water activity levels are necessary before spoilage is likely to occur. This kind of spoilage may affect products with hydrophilic and deliquescent components, or products subjected to surface condensation due to temperature or pressure fluctuations. Once growth has begun, metabolically produced water may escalate deterioration (Beveridge 1992).

8.3.3.2 Yeasts

The water requirements of yeasts are not as well documented as those for other fungi, but the review by Scott (1957) reports agreement in the literature that yeasts have a somewhat higher water requirements than moulds. Studies by Burcik (1950) of 14 yeast strains stored at humidities controlled by sodium chloride solutions indicate a minimum a_w for growth which varied only between 0.88 and 0.91. There are of course strains of osmophilic yeasts which are known to grow at much lower a_w. von Schelhorn (1950) showed the remarkable resistance of *Zygosaccharomyces barkeri* (a strain of *Saccharomyces rouxii*) which can grow slowly in fructose syrup at $0.62a_w$. However, since the growth of osmophilic strains is highly dependent upon the total water content, and the presence of specific types of solute in the substrate solution, they are not a problem encountered with solid dosage forms produced from dry powders.

For yeasts in general, the limiting value of a_w for growth is a characteristic of the particular strain, which is resistant to change even through repeated subculturing. Scott (1936) reported that, as is the case for moulds, the germination of yeast spores will show an increase in lag time and a reduction in the rate of growth when a_w is reduced.

8.3.3.3 Bacteria

Troller and Christian (1978) published minimum tolerated water activities for the growth of some species of bacteria. For species of interest, a_w minima varied between 0.86 for *Staph. aureus* to 0.97 for a strain of *Clostridium botulinum*. Growth at the a_w minimum was in each case recorded when all other conditions (temperature, pH, oxygen tension, nutrients) were optimum. 'Growth' is taken as spore germination, or an extremely slow increase in the numbers of viable organisms. There are halophilic species of bacteria which are tolerant of low a_w, and incapable of growing at high a_w. The growth of halophilic bacteria can only take place in the presence of specific inorganic salts, of which NaCl is the most important. Halophilic organisms however present no significant spoilage threat.

Most (non-halophilic) bacteria have an optimum a_w for growth in the region of 0.980 to $0.997a_w$ (Troller and Christian 1978). When grown within this a_w range, bacteria are extremely sensitive to small changes in a_w. It has been suggested that more dilute laboratory media (for example nutrient broth, which has an a_w of 0.999) may provide suboptimal conditions for growth (Christian and Waltho 1962). Troller and Christian (1978) reported that the Gram-negative species are most sensitive to a_w, with minima for growth in the range of 0.96 to $0.94a_w$. *Bacillus* species will grow at water activities between 0.93 and $0.90a_w$. The widest a_w tolerance is shown by the Gram-positive cocci, with minima between 0.95 and $0.83a_w$. Bacterial growth curves show increased lag time as the a_w decreases away from optimum; under the same conditions the growth rate decreases, and a plot of growth rate versus a_w will give a straight line of negative slope.

The minimum a_w for enterotoxin production by strains of *Staph. aureus* is higher than the minimum water requirement for growth (Troller and Stinson 1975). Any reduction in the environmental a_w, away from optimum, causes a dramatic reduction in enterotoxin output. This may explain why significant numbers of these organisms may be isolated from spoiled products, with no detectable enterotoxin. The staphylococci and micrococci are poor competitors, and are rapidly overgrown by other organisms when grown in moist conditions. Their relative importance in food spoilage is due to their ability to grow in quite dry or concentrated environments. Once microbial growth has been initiated by such organisms, metabolically produced moisture will increase local a_w levels, further accelerating growth and allowing previously inhibited organisms to grow (Mossel 1975, Beveridge 1992).

Studies of vegetative bacteria in powders and tablets over the entire humidity range have generally shown growth only at 100% RH and just below, with the best chance of organisms surviving at low relative humidities. In a_w conditions that are too dry or concentrated for growth, a_w is still seen as an important factor in determining the survival of a bacterial population of vegetative cells or spores. For spores and dried vegetative cells stored at a range of relative humidities, survival was good between 5% and 20% RH (Higginbottom 1953). In this study, it was observed that vegetative cells and spores died below 5% RH, and storage at 80–100% RH resulted in rapid bacterial growth followed by overgrowth by moulds. More detailed studies have been conducted in our own laboratories (Blair *et al.* 1987, Blair 1989). Tablets were produced from excipients contaminated with vegetative bacteria, and viable counts monitored during storage at a range of relative humidities. Whilst *Enterobacter cloacae* (a spoilage isolate), and to some extent *Staph. aureus* (standard strain), proliferated in tablets at 100% RH, at 95% RH and below, the numbers of viable organisms generally fell over 2–4 days at 25°C. The death rates were greatest at humidities close to 95% RH, reflecting poor tolerance of the organisms to sub-optimal growth conditions, and highest for *Staph. aureus*, which has more exacting nutritional requirements than *Ent. cloacae*. The survival of organisms was best at the lowest humidities studied, 23% and 46%, where there was too little water to initiate growth. For non-proliferating organisms at relative humidities below 80%, patterns of survival varied according to the a_w, and survival was better for organisms in more hygroscopic excipients, such as potato starch.

At equivalent water activities, the characteristic surface relationship between excipient and adsorbed water may exert its own effect on the contaminating organisms adhering to the solid, so that equation (8.1) does not fully account for water availability.

8.3.4 *The formulator and water activity*

The conditions required for the survival of yeasts, moulds and bacteria are completely different from those required for rapid growth. It is not surprising that many organisms survive in solid products. However, spoilage due to microbial growth is more likely to be due to moulds, which can grow in conditions of lower a_w than non-osmophilic yeasts or bacteria.

It should be within the formulator's power to use a knowledge of microbial behaviour at different a_w levels, together with the effect of formulation factors on the water distribution within a solid product, to develop a self-preserving system. Solid formulations of a hydrophobic basis are least likely to allow local high a_w for microbial growth, or a suitable moisture content to support microbial survival. Substances which lower water activity (e.g. polyethylene glycols) will cause any localized solutions formed on the product at high humidity to be disfavourable to growth. Conversely, at relative humidities below 80%, even non-spore-forming bacteria can survive in apparently dry products, if the surface–water relationship is favourable, e.g. in hygroscopic excipients such as starches.

It is unlikely that preservation considerations will be a primary concern for solid dosage forms. However, for products developed for international markets where high humidity must be considered, formulation to achieve preservation could be a cheaper and logistically simpler option than using moisture-protective packaging. Where alternative excipient combinations offer the same processing benefits, the formulation with a low hygroscopicity may well be better preserved. The use of antimicrobial preservatives in such systems is discussed below (section 8.5).

8.4 The effect of manufacturing methods on the microbial quality of solid dosage forms

The likelihood that microbial contaminants in dry raw materials will persist into the finished products, will depend on the water activity to which the materials are subjected, and the processing variables used. With a view to different effects on micro-organisms, tablet production may take one of two general routes: wet granulation involves the introduction of an aqueous phase and subsequent drying: dry methods, such as direct compression, do not. In their simplest form, hard gelatin capsules may be filled with blended dry excipients and drug.

8.4.1 *Granulation and drying*

Wet granulation is a process which involves moistening powder blends and drying the wet mass at temperatures generally not in excess of 50°C. The purpose of this process is to produce free-flowing and compressible powder agglomerates. Such conditions might well be expected to encourage microbial growth. Fassihi and Parker (1977) reported that granule drying (at 40–45°C) resulted in the death of bacterial cells in 12 hours, and 99% of mould spores in 24 hours. This report is in keeping with the experience of industrial formulation scientists (personal communication, Long 1986) and with the work of Wallhausser (1977). It is therefore likely that contaminated powders processed by wet granulation will be presented for tableting with at worst a very low load of spores.

For the purposes of tableting, most pharmaceutical drying processes are designed such that some physically bound water remains in the 'dried' granules. It is often undesirable, uneconomic and unnecessary to raise the solids above the wet bulb temperature (Ganderton 1968, Lachman *et al.* 1976). However, it may be that this practice permits survival of microbial contaminants, especially spores. Bacterial spores were shown to be most resistant to thermal inactivation on drying when they are almost, but not completely dry (Murrell and Scott 1966). These authors reported that the optimum a_w for spore survival was between $0.20a_w$ and $0.40a_w$, depending on the bacterial species. On spray drying, micro-organisms are protected from high temperatures if the droplet size is large, and dehydration therefore slow (Loncin *et al.* 1968). The droplet temperature starts to rise more sharply when the moisture content reaches a low level, by which point the contaminating organisms are more resistant to heat. For contaminated excipients, Blair (1989) confirmed that on drying, the extent of bacterial inactivation depended on the rate of water loss. The greatest inactivation was recorded for excipients with a low equilibrium moisture content, which lose physically bound water most rapidly. It has been shown that micro-organisms are afforded a further degree of protection by the presence of organic matter in the drying vehicle (Ishag 1973). A classical text on the survival of bacteria during and after drying is that of Fry and Greaves (1951), to which the reader may make further reference.

There have been comparative studies of microbial counts in finished tablets after wet and dry granulation. Chesworth *et al.* (1977) found that viable counts in tablets were reduced by 70% compared with the compression mix, by both wet granulation and drying, or direct compression. Ibrahim and Olurinola (1991) found that tablets produced by direct compression had lower viable counts than those produced by wet granulation. In this case, contaminated gelatin had been used as the binder for the wet granulation; the viable count was reduced almost tenfold on drying at 50°C, but the mixture was recontaminated by using talc containing nearly 1000 organisms per gram as the lubricant.

8.4.2 Compaction

Contaminating micro-organisms may be destroyed during tablet compression. Significant factors in determining the extent of kill during compaction would appear to be the compaction pressure, speed, the nature of the formulation and the distribution of the micro-organisms within the formulation.

The effect of compaction pressure on the survival of fungal spores was shown to be a first-order profile, with 100% kill being achieved when a mean compaction pressure of $27 \, MN^{-2}$ was applied (Fassihi *et al.* 1977). These authors suggested that local heat production contributed to the level of inactivation. Conversely, Yanagita *et al.* (1978) observed a non-linear relationship between compression and survival of *E. coli* cells compressed in a skimmed milk base with microcrystalline cellulose. This non-linear kill is thought to be due to the transmission of lethal shear forces through the compressed powders. Plumpton *et al.* (1982) reported the survival of micro-organisms in various compression materials used for tableting, and noted that survival was dependent on the nature of the compression material, and the degree of kill was directly proportional to the size of the contaminating cells. They argued that micro-organisms are killed during tablet compression by shearing forces, and that temperature increases are not a significant factor. Increasing compaction speed was found to reduce the level of kill achieved, and this was thought to be due to stress relaxation, generating greater shear forces at slower speeds. Larger cells were more susceptible to kill than smaller ones, which may be explained by the greater effect of shear on larger cells. For materials which fracture on compression, kill was proportional to applied force over the whole pressure range, whereas for more plastically deforming materials, a decrease in the ratio of applied pressure to per cent kill was observed as minimum porosity was approached. Plumpton *et al.* (1986) reported that the inactivation of *Bacillus megaterium* spores in tablet formulations, varied according to whether the spores were added in a dry compression vehicle, in a granulation, or in granulating fluid. The authors explained these observations in terms of spatial protection or exposure to the likely sites of high shear in the particular compression system.

Blair *et al.* (1991) compared compression/survival plots for three excipients with different compaction mechanisms, which had been contaminated with *Staph. aureus* (Fig. 8.1). Lactose monohydrate is a brittle crystalline material, and tends to consolidate by fracture initiated at the early stages of compression. Fig. 8.1 shows that bacterial survival is inversely proportional to applied pressure, probably resulting from increasing the areas of interparticulate contact on fracturing, so that the cells are inactivated by shearing forces as described by Plumpton *et al.* (1986). For maize starch, which is prone to plastic flow, shear forces are thought to be magnified by plastic deformation accompanied by particle rearrangement in the early stages of compression, causing extensive inactivation of contaminants. As the compression force is increased however, particle rearrangement is hindered by densification of the starch compact, so the extent of kill is reduced. This pattern can be seen on the graph. The compression/survival plot for contaminated microcrystalline cellulose, initially follows the same pattern as for maize starch. However, at compression forces exceeding 8 kN (corresponding to 92 MPa for the punches used), a linear relationship is seen. The authors explain that microcrystalline cellulose undergoes

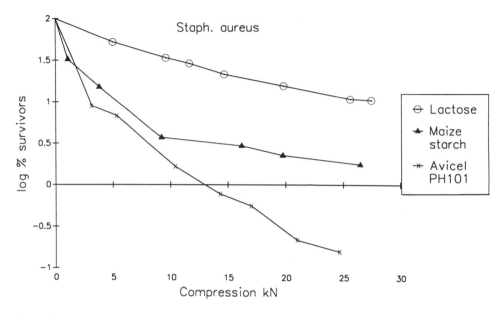

Figure 8.1 Compression/survival plots for three excipients, with different compaction mechanisms, contaminated with *Staph. aureus*. (Reproduced, with permission, from Blair *et al.*, 1991)

plastic deformation, like maize starch, but that above 80 mPa, the material exceeds its plastic limit, and at higher pressures consolidates by brittle fracture. This explains the change in shape of the graph.

Further evidence of the relative importance of shear forces in the inactivation of micro-organisms has been provided by studies of bacterial inactivation during ram extrusion of a contaminated formulation (Blair 1989). Measured compression forces were applied to a column of wet powder mass, contaminated with *Ent. cloacae*. Material was sampled from the compressed column and from the extrudate. It was found that the highest levels of bacterial inactivation were consistent with the samples which had been subjected to the highest shear stresses (against the wall of the barrel and die).

8.4.3 *Gelatin capsules*

Gelatin is one of the named non-sterile materials for which the current *British Pharmacopoeia* limit tests stipulate the absence of *E. coli*, *Salmonella* species and *Pseudomonas* species. In surveys of pharmaceutical raw materials, only low levels of non-pathogenic microbial contaminants have been reported (Kruger 1973, Pedersen and Ulrich 1968). Van Doorne and Boer (1987) showed that hand-filled paracetamol capsules from pharmacy samples, were contaminated with low levels of *Staph. aureus* (below the *USP* limit). The contamination had arisen because of insufficient hygienic precautions during filling.

Soft gelatin capsules contain product formulated in an oil, semi-solid or semi-organic basis, which are unlikely to be favourable for microbial growth. Hard gelatin capsules undergo drying during their manufacture which leaves them with a low microbiological load (Garcia-Arribas *et al.* 1986). Problems associated with frank microbial growth are not encountered since this would require local water activities significantly higher than the upper limits for physical preservation of gelatin (about 70% RH). Consequently, hard gelatin capsules would have physically deteriorated before growth of contaminating micro-organisms could begin. Good manufacturing practice on filling and packing, and suitable storage conditions, should therefore prevent any risk of spoilage.

8.5 Preservation of solid pharmaceuticals

From section 8.4, it can be seen that tablets manufactured under GMP conditions will at worst contain only very low levels of contamination at the end of production. The preservation of solid formulated products then depends on the nature of the formulation (in terms of water access) and the environment into which it will pass (in terms of potential contamination and relative humidity). Thus, Waterman *et al.* (1973) observed the complete inability of *Staph. aureus* to survive on tested solid dosage forms, e.g. a barbiturate which is hydrophobic in nature, and aspirin, which produces an aqueous dispersion of low pH. Other products in this study showed lower levels of survival than the control, demonstrating perhaps that these formulations had some natural tendency towards self-preservation.

The inclusion of antimicrobial preservatives into finished solid dosage forms has been investigated. Smart and Spooner (1972) found that the inclusion of 0.1% w/w propyl *p*-hydroybenzoate in Paracetamol tablets B.P. slowed and inhibited the growth of moulds, under conditions of high humidity. For tablets prepared from lactose and potato starch, the inclusion of 0.0 5% to 0. % w/w parabens (methyl, ethyl or propyl *p*-hydroxybenzoic acid) was found to protect against subsequent fungal challenge (Fassihi *et al.* 9 8). It was found that tablet granules which had been sprayed externally with an aqueous or alcoholic solution of preservative, were better protected than tablets where the preservative was included in the granulation. Bos *et al.* (989) reported that inclusion of % w/w sodium methyl *p*-hydroxybenzoate or potassium sorbate in the binder solution used for starch tablets was effective against challenge with *Aspergillus niger*, and prevented growth of natural contaminants on storage at tropical conditions. Bos *et al.* (99) advocated the inclusion of % w/w sodium methyl *p*-hydroxybenzoate to optimize tablet formulations for use in tropical countries.

Aqueous film coating procedures use solutions of cellulosic polymers which may readily support microbial growth (Banker *et al.* 98). Where large batches of solution are prepared in advance, preservatives may be added to prevent subsequent microbial proliferation in pumps, lines and coating equipment. The results of Fassihi *et al.* (9 8) suggest that a preserved film coating could protect finished tablets from possible microbial growth during storage.

8.6 Concluding comments

The subject of microbial contamination in solid dosage forms deserves serious consideration. Many solid dosage forms are entirely self-preserving either in chemical nature, or by virtue of their dry state. However, there have been serious incidences of contamination, justifying compendial control.

As yet, microbiologists do not have a complete understanding of the factors which determine the growth and survival of pathogenic and spoilage micro-organisms in solid dosage forms. The studies reported to date have used a variety of methods for contaminating products and raw materials, which may be the source of misleading results. Ishag (1973) showed that micro-organisms grown in excipient powders are more resistant to inactivation than micro-organisms inoculated into a powder from another growth medium. Viable counting methods applied to dry products generally involve dispersion of the test samples in a nutrient liquid. Such methods may give poor recovery of viable organisms compared with slow rehydration (Kosanke *et al.* 1992), and hence conventional viable counts may well underestimate contamination levels.

Preservation of solid dosage forms can be achieved by using raw materials of suitable quality and by adjusting and understanding, formulation and moisture effects. For this reason, the routine use of preservatives in solid oral dosage forms is unlikely to be worthwhile, or justifiable to licensing authorities.

Since the introduction and adoption of microbial limits and GMP for solid oral dosage forms, it is generally considered that such products do not present a significant microbiological hazard. However, microbiologists must be aware of new developments in formulation technology, since micro-organisms can adapt to new growth opportunities, and opportunist organisms can flourish in apparently hostile conditions. For example, *Listeria* species became relatively important pathogens in the food industry, by virtue of their ability to grow at refrigeration temperatures. Micro-organisms can find a niche even in extreme environments, and the failure to anticipate situations which may encourage survival and growth of contaminants could have serious consequences.

References

AHLNECK, C., and ZOGRAFI, G. (1990) The molecular basis of moisture effects on the physical and chemical stability of drugs in the solid state. *Int. J. Pharm.* **62** 87–95.

ANDERSON, R. B. (1964) Modifications of the Brunauer, Emmett and Teller equation. *J. Am. Chem. Soc.* **68** 686–691.

ANON. (1993a) *British Pharmacopoeia.* HMSO, London.

(1993b) International conference on harmonisation of technical requirements for registration of pharmaceuticals for human use. Orlando, FL: The tripartite guideline for the stability testing of new drug substances and products. Commission of the European Communities Directorate—General Industry. Document 111/3335/92-EN, draft no. 4.

(1995) *European Pharmacopoeia*, 2nd edn, VIII 15. Maison Neuve, Sainte Ruffine.

(1995) *United StatePharmacopoeia.* XXIII. The United States Pharmacopoeial Convention Inc., Rockville, MD.

AYERST, G. (1969) The effect of water and temperature on growth and spore germination of some fungi. *J. Stored Prod. Res.* **5** 127–141.

BANKER, G., PECK, G., WILLIAMS, E., TAYLOR, D., and PIRAKITIKULR, P. (1982) Microbiological considerations of solutions used in aqueous film coating. *Drug. Dev. Ind. Pharm.* **8**(1) 41–51.

BEVERIDGE, E. G. (1992) In: Hugo, W. B., and Russell, A. D. (eds), *Pharmaceutical Microbiology.* 5th edn. Blackwell Scientific, Oxford, p. 379.

BHASKAR, G., FORD, J. L., HOLLINGSBEE, D. A., and EDWARDSON, P. A. D. (1993) Characterisation of water uptake by hydrocolloids using differential scanning calorimetry. *J. Pharm. Pharmacol.*, Suppl. 80P.

BIGGER, J. W., and NELSON, J. H. (1941) The growth of coliform bacilli in distilled water. *J. Path. Bact.* **53** 189–206.

BLAIR, T. C. (1989) *Some factors influencing the survival of microbial contamination in solid oral dosage forms.* Ph.D. Thesis, King's College, University of London.

BLAIR, T. C., BUCKTON, G., and BLOOMFIELD, S. F. (1987) Water available to *Enterobacter cloacae* contaminating tablets stored at high relative humidities. *J. Pharm. Pharmacol.* **39** (suppl.) 125P.

BLAIR, T. C., BUCKTON, G., BEEZER, A. E., and BLOOMFIELD, S. F. (1990) The interaction of various types of microcrystalline cellulose and starch with water. *In. J. Pharm.* **63** 251–257.

BLAIR, T. C., BUCKTON, G., and BLOOMFIELD, S. F. (1991) On the mechanism of kill of microbial contaminants during tablet compression. *Int. J. Pharm.* **72**(2) 111–115.

BONNER, J. T. (1948) A study of the temperature and humidity requirements of *Aspergillus niger*. *Mycologia* **40** 728–738.

BOS, C. E., VAN DOORNE, H., and LERK, C. F. (1989) Microbial stability of tablets stored under tropical conditions. *Int. J. Pharm.* **55** 175–183.

BOS, C. E., BOLHUIS, G. K., and LERK, C. F. (1991) Optimization of tablet formulations based on starch/lactose granulations for use in tropical countries. *Drug Dev. Ind. Pharm.* **17**(17) 2373–2389.

BOWMAN, F. W., WHITE, M., and LYLES, R. I. (1971) Microbial contamination of non-sterile antibiotic market samples: a survey. *J. Pharm. Sci.* **60**(7) 1099–1101.

BRUNAUER, S., EMMETT, P. H., and TELLER, E. (1938) Adsorption of gases in multimolecular layers. *J. Am. Chem. Soc.* **60** 309–319.

BUCK, A. C., and COOKE, E. M. (1969) The fate of ingested *Pseudomonas aeruginosa* in normal persons. *J. Med. Microbiol.* **2** 521–525.

BUCKTON, G., BEEZER, A. E., and NEWTON, J. M. (1986) A vacuum microbalance technique for studies on the wettability of powders. *J. Pharm. Pharmacol.* **38** 713–720.

BURCIK, E. (1950) Uber die Beziehungen swischen Hydratur und Wachstum bei Bakterien un Hefen. *Arch. Mikrobiol.* **15** 203–235.

CENTER FOR DISEASE CONTROL (1966). *Salmonella surveillance*, Annual report of the Center for Disease Control. Center for Disease Control, Atlanta, GA, pp. 1–16.

CHESWORTH, K. A. C., SINCLAIR, A., STRETTON, R. J., and HAYES, W. P. (1977) The effect of tablet compression on the microbial content of granule ingredients. *Microbios. Lett.* **4** 41–45.

CHRISTIAN, J. H. B., and WALTHO, J. (1962) The water relations of staphylococci and micrococci. *J. Appl. Bact.* **25**(3) 369–377.

COYLE, E. F., RIBEIRO, C. D., HOWARD, A. J., PALMER, S. R., JONES, H. I., WARD, L., and ROWE, B. (1988). *Salmonella enteriditis* phage type 4 infection associated with hens' eggs. *The Lancet* **ii** 1295–1297.

D'AOUST, J. Y. (1985) Infective dose of *Salmonella typhimurium* in Cheddar cheese—brief report. *Am. J. Epidemiol.* **122** 717–720.

DE BOER, J. H. (1968) *The dynamic character of adsorption*, 2nd edn. Clarendon Press, Oxford.

FARRAR, W. E. (1963) Serious infection due to 'non-pathogenic' organisms of the genus *Bacillus*. *Am. J. Med.* **34**(1) 135–141.

FASSIHI, A. R., and PARKER, M. S. (1977) The influence of water activity and oxygen tension upon the survival of *Aspergillus* and *Penicillium* species on tablets. *Int. Biodeter. Bull.* **13** 75–80.

FASSIHI, A. R., DAVIES, P. J., and PARKER, M. S. (1977) The effect of punch pressure on the survival of fungal spores during the preparation of tablets from contaminated raw materials. *Zentralbl. Pharm. Pharmakother. Laboratoriumsdiag.* **116** 1267–1271.

FASSIHI, A. R., PARKER, M. S., and DINGWALL, D. (1978) The preservation of tablets against microbial spoilage. *Drug Dev. Ind. Pharm.* **4** 515–527.

FEDERATION INTERNATIONALE DE PHARMACEUTIQUE (1975) *Pharm. Acta Helv.* **50** 125–131.

FERGUSON, W. W., and JUNE, R. C. (1952) Experiments on feeding adult volunteers with *Escherichia coli* 111, B4, a coliform organism associated with infant diarrhoea. *Am. J. Hyg.* **55** 155–169.

FISCHER, A., FUGLSANG-SMIDT, B. and ULRICH, K. (1968) Microbial content in non-sterile pharmaceuticals IV: Tablets. *Dan. Tidsskr. Farm.* **42** 125–131.

FRY, R. M., and GREAVES, R. I. N. (1951) The survival of bacteria during and after drying. *J. Hyg.* **49** 220–246.

FUGLSANG-SMIDT, B., and ULRICH, K. (1968) Microbial content in non-sterile pharmaceuticals I. *Dan. Tidsskr. Farm.* **42** 257–263.

GALLOWAY, L. D. (1935) The moisture requirements of mold fungi with special reference to mildew in textiles. *J. Text. Inst.* **26** T123–T129.

GANDERTON, D. (1968) In: *Unit processes in pharmacy.* Morrison Gibb, London, pp. 89–115.

GARCIA-ARRIBAS, M. L., MOSSO, M. A., DE LA ROSA, M. C., and GASTON DE IRIARTE, E. (1983) *Cienc. Ind. Farm. CIDFA,* **2** 133–137.

GARCIA-ARRIBAS, M. L., DE LA ROSA, M. C., and MOSSO, M. A. (1986) Characterisation des souches de Bacillus isolees a partir de medicaments solides administres par voie orale. *Pharm. Acta Helv.* **61** 303–307.

GARCIA-ARRIBAS, M. L., IAZA, C. J., DE LA ROSA, M. C. and MOSSO, M. A. (1988) Characterisation of *Bacillus cereus* strains isolated from drugs and evaluation of their toxins. *J. Appl. Bact.* **64**(3) 257–264.

GILL, O. N., BARTLETT, C. L. R., SOCKETT, P. N., and VAILE, M. S. B. (1983) OUTBREAK of *Salmonella napoli* infection caused by contaminated chocolate bars. *The Lancet* **i** 575.

GREENWOOD, M. H., and HOOPER, W. L. (1983) Chocolate bars contaminated with *Salmonella napoli. Br. Med. J.* **286** 1384.

GUGGENHEIM, E. A. (1966) *Applications of statistical mechanics,* 2nd edn. Clarendon Press, Oxford.

HEINTZELER, I. (1939) Das Wachstum der Schimmelpilze in Abhangigkeit von der Hydraturverhaltnissen unter verschiedenen Aussenbedingugen. *Arch. Mikrobiol.* **10** 92–132.

HIGGINBOTTOM, C. (1953) The effect of storage at different relative humidities on the survival of micro-organisms in milk powder and in pure cultures dried in milk. *J. Dairy Res.* **20** 65–75.

IBRAHIM, Y. K. E., and OLURINOLA, P. F. (1991) Comparative microbial contamination levels in wet granulation and direct compression methods of tablet preparation. *Pharm. Acta Helv.* **66**(11) 298–301.

ISHAG, A. H. O. (1973) *Studies on some factors affecting the viability of bacteria in powders of pharmaceutical interest,* Ph.D. Thesis, University of London.

JAIN, N. K., and CHAUHAN, C. S., (1978) Microbiological contamination in antacid tablets. *Indian J. Hosp. Pharm.* September–October 137–138.

KALLINGS, L. O., ERNERFELDT, F., and SILVERSTOLPE, L. (1966a) *Microbiological contamination of medical preparations,* Report 1965 to the Swedish National Board of Health; Transactions of the Swedish Medical Society 1965. *Acta Path. Microbiol. Scand.* **66** 287.

KALLINGS, L. O., RINGERTZ, O., SILVERSTOLPE, L., and ERNERFELDT, F. (1966b) Microbiological contamination of medical preparations. *Acta Pharm. Suec.* **3** 219–228.

KHANTE, S., NIKORE, R. L., and JOSHI, S. B. (1979) Microbial contamination studies in sterile and non-sterile pharmaceutical formulations in consumers' storage conditions. *Indian J. Hosp. Pharm.* **16**(4) 114–117.

KOSANKE, J. W., OSBURN, R. M., SHUPPE, G. I., and SMITH, R. S. (1992) Slow rehydration improves the recovery of dried bacterial populations. *Can. J. Microbiol.* **38**(6) 520–525.

KRUGER, D. (1973) A contribution to the subject of microbial contamination of active ingredients and adjuvants. *Drugs Made Ger.* **16** 109–125.

LACHMAN, L., LIBERMAN, H. A., and KANIG, J. L. (1976) In: *The theory and practice of industrial pharmacy.* 2nd edn. Lea and Febiger, Philadelphia, Pennsylvania, pp. 503–524.

LOCIN, M., BIMBINET, J. J., and LENGES, J. (1968) Influence of the activity of water on the spoilage of foodstuffs. *J.Fd. Technol.* **3** 131–142.

LONG, D. (1986) Personal communication. Glaxo Group Research, Ware, Hertfordshire.

McCULLOUGH, N. B., and EISELE, C. W. (1951) Experimental human salmonellosis I. Pathogenicity of strains of *Salmonella meleagridis* and *Salmonella anatum* obtained from spray-dried whole egg. *J. Infect. Dis.* **88** 278–289.

MOSSEL, D. A. A. (1975) Water and micro-organisms in foods—a synthesis. In: Duckworth, R. B. (ed.), *Water relations of foods.* Academic Press, London.

MURREL, W. G., and SCOTT, W. J. (1966) The heat resistance of bacterial spores at various water activities. *J. Gen. Microbiol.* **43** 411–425.

NANDAPURKAR, S. N., SHANKHPAL, K. V., and DEOBADE, N. G. (1985) Bacteria isolated from the pharmaceutical preparation. Part I. Tablets. *Ind. J. Hosp. Pharm.* **23** 131–139.

PEDERSEN, A., and ULRICH, K. (1968) Microbial content in non-sterile pharmaceuticals. III. Raw materials. *Dan. Tidsskr. Farm.* **42** 71–83.

PITT, J. I. (1975) In: Duckworth, R. B. (ed.), *Water relations of food.* Academic Press, London, pp. 273–307.

PLUMPTON, E. J., FELL, J. T., and GILBERT, P. (1982) Survival of microorganisms during compaction in various direct compression materials used for tableting. *Microbiols. Lett.* **21** 7–15.

PLUMPTON, E. J., GILBERT, P., and FELL, J. T. (1986) Effect of spatial distribution of contaminant microorganisms within tablet formulations on subsequent inactivation through compaction. *Int. J. Pharm.* **30** 237–240.

SCHILLER, I., KUNTSCHER, H., WOLFE, A., and NEKOLA, M. (1968) Microbial content of non-sterile therapeutic agents containing natural or semi-natural active ingredients. *Appl. Microbiol.* **16** 1924–1928.

SCHNELLER, G. H. (1978) Microbial testing of oral solid dosage forms. *Drug. Cosmet. Ind.* **122** 48 et seq.

SCOTT, W. J. (1936) The growth of microorganisms on Ox muscle. I. The influence of water content of substrate on rate of growth at –1°C. *J. Counc. Sci. Ind. Res. Aust.* **9** 177–190.

(1957) Water relations of food spoilage microorganisms. *Adv. Food. Res.* **7** 83–127.

SHERIDAN, P. L., BUCKTON, G., and STOREY, D. E. (1993) Comparison of two techniques for the assessment of the wettability of powders. *J. Pharm. Pharmacol.* (suppl.) 1P.

SMART, R., and SPOONER, D. (1972) Microbial spoilage in pharmaceuticals and cosmetics. *J. Soc. Cosm. Chem.* **23** 721–737.

SOMERVILLE, P. C. (1981) A survey into microbial contamination of non-sterile pharmaceutical products. *Farm. Tijdschr. Belg.* **58** 345–350.

TOMKINS, R. G. (1929) Studies of the growth of molds. *Proc. R. Soc. B* **105** 375–401.

TROLLER, J. A., and CHRISTIAN, J. H. B. (1978) In: *Water activity and food.* Troller, J. A., and Christian, J. H. B. (eds). Academic Press, New York.

TROLLER, J. A., and STINSON, J. V. (1975) Influence of water activity on growth and enterotoxin formation by *Staphylococcus aureus* in foods. *J. Fd. Sci.* **40** 802–804.

TURNBULL, P. C. B., JORGENSEN, K., KRAMER, J. M., GILBERT, R. J., and PARRY, J. M. (1979) Severe clinical conditions associated with *Bacillus cereus* and the apparent involvement of exotoxins. *J. Clin. Path.* **32** 289–293.

VAN DEN BERG, C., and BRUIN, S. (1981) Water activity and estimation in food systems. In: *Water activity: influences on food quality.* Rockland, L. B., and Stewart, G. F. (eds). Academic Press, New York.

VAN DOORNE, H., and BOER, Y. (1987) Microbiological quality of hand-filled gelatin capsules. *Pharm. Weekbl.* **122** 820–823.

VON SCHELHORN, M. (1950) Untersuchengen uber den Verberb wasserarmer Lebensmittel durch osmophile Mikroorganismen. I. Verberb von Lebensmittel durch osmophile Hefen. *Z. Lebensm.-Unters. Forsch.* **91** 117–124.

WALLHAUSER, K. H. (1977) Microbial aspects on the subject of oral solid dosage forms. *Pharm. Ind.* **39** 491–497.

WATERMAN, R. F., SUMNER, E. D., BALDWIN, J. N., and WARREN, F. W. (1973) Survival of *Staphylococcus aureus* on pharmaceutical oral solid dosage forms. *J. Pharm. Sci.* **62** 1317–1320.

WHITE, M., BOWMAN, F. W., and KIRSCHBAUM, A. (1968) Bacterial contamination of some non-sterile antibiotic drugs. *J. Pharm. Sci.* **57** 1061–1063.

WOZNIAK, W. (1971) Studies on microbiological purity of plant oral medicines. *Acta Polon. Pharm.* **28**(1) 129–135.

YANAGITA, T., MIKI, T., SAKAI, T., and HORIKOSHI, I. (1978) Microbial studies on drugs and their raw materials. I. Experiments on the reduction of microbial contaminations in tablets during processing. *Chem . Pharm. Bull.* **26** 185–190.

9

Development of preservative systems

S. P. DENYER

Department of Pharmacy, University of Brighton, Moulsecoomb, Brighton BN2 4GJ

Summary

(1) Complex pharmaceutical, cosmetic and toiletry formulations are quite frequently protected from microbial contamination by multicomponent *preservative combinations*.

(2) A minimum requirement of any combination is that it achieves at least the same level of protection overall as the summed effect of the individual components. Antagonistic mixtures are to be avoided.

(3) The antimicrobial outcome of an interaction between preservative agents depends upon established biochemical and biophysical principles.

(4) In order to confirm the suitability, and to determine the optimum composition, of promising preservative combinations *in vitro* tests should be performed. In the final analysis, the preservative system must be evaluated in the intended product.

(5) The success of a preservative combination is dependent upon a rational selection of individual components; this should be based upon a sound understanding of preservative action and behaviour.

9.1   Introduction

Currently, the available 'pool' of acceptable preservative agents for pharmaceutical and cosmetic use, if not actually contracting, is certainly showing no signs of increasing (Denyer and Wallhaeusser 1990). This is partly due to the high costs involved in developing and testing new compounds of acceptable selective toxicity and also partly to an increased anxiety over the safety of some existing agents (Bloomfield 1986). To the formulator of pharmaceutical, cosmetic or toiletry products, therefore, is often left the difficult task of achieving adequate preservation with an inadequate range of preservative agents. Thus, there is a need to understand and exploit the properties of existing agents to their fullest and, in this chapter, this philosophy is extended to a consideration of their use in combination.

Successful preservation, especially of a complex formulation, can sometimes be elusive. Many factors contrive to oppose the preservative agent. These include the physicochemical characteristics of the formulation, the breadth, level and type of microbial challenge, possible interactions with formulation ingredients, and preservative instability. Not surprisingly, a single agent is often unable to preserve a complex product adequately and, increasingly, a formulator is required to employ a *multi-component preservative system* in an attempt to provide the necessary antimicrobial cover. It is axiomatic that such a system must confer some benefit over the individual preservative agents from which it is comprised. Thus, it is necessary, and it should be possible, to develop a preservative system which is optimal for its particular application and micro-biological challenge; this may be achieved by a careful consideration of the antimicrobial action and behaviour of individual components. In this way, a set of principles relevant to the selection of compatible preservative partners may be established.

9.2 The use of preservative combinations

Significant practical benefits and advantages can be gained by the use of preservative combinations. These have been considered by Crowshaw (1977) to include the following:

(1) an increased spectrum of activity;

(2) the use of lower concentrations of individual components, resulting in a possible reduction in toxicity;

(3) the prevention of the development of microbial resistance to individual preservatives;

(4) a possible overall enhancement of antimicrobial activity, beyond that expected from simple addition (*synergy*);

(5) an extended time course of preservation achieved by combining a labile, markedly biocidal preservative with a stable longer-acting agent.

Furthermore, careful selection of individual agents for combination, based on their physicochemical properties, may serve to overcome microbiological problems created by the physical limitations of individual preservative agents. In this respect, the factors affecting preservative efficacy must be carefully considered (Denyer and Wallhaeusser 1990, Van Doorne 1990).

A minimum requirement of any combination should be that it achieves at least the same level of protection overall as the sum of the individual components. Ill-considered preservative combinations may lead to the inclusion of irrelevant agents (Parker 1982) or, worse, may reveal *antagonism*, either between themselves (Rehm 1959, Rehm and Stahl 1959, 1960, Pons *et al.* 1992) or, individually, with components of the formulation (Richards and Reary 1972, Richards and McBride 1973, Farouk 1981, McCarthy 1984). The basis of any rational approach to the use of preservative combinations must be therefore to consider only those agents offering antimicrobial *additivity* or synergy. It is, for instance, counter-productive to select agents solely on the basis of broadening the spectrum of activity in combination if they are subsequently shown to compromise efficacy against a particular contaminant by antagonism. In order to avoid or reduce the possibility of antagonism, attention should be paid to the *mechanism of antimicrobial action* before selection of components. At worse, this might ensure a compatible additive system and at best it may achieve a measure of antimicrobial synergy.

9.3   Mechanisms of action: prediction of enhanced activity

There is a growing awareness of the need to establish mechanisms of action for antimicrobial agents. This information assists in the design of new compounds and in the understanding of resistance mechanisms and provides a focus for toxicological attention, as well as offering a possible basis for predicting the effects of combination. Certainly, mechanism-of-action studies (Harold 1970, Franklin and Snow 1981, Hugo 1982, Denyer 1990, Denyer and Hugo 1991) have shown that preservative agents can no longer be considered as general cell poisons and thus their activity may be optimized by design; a summary of likely targets for a range of preservative agents is given in Fig. 9.1.

When acting simultaneously on a uniform microbial population, combinations of antimicrobial agents may cause variously an increased, decreased or unchanged antimicrobial response when compared with their summed individual effects. In general, those agents from the same chemical group or having the same mode of action are likely to produce merely additive effects while those exhibiting different mechanisms or sites of action may either serve to reinforce (synergize) or reduce (antagonize) their individual activities. Undoubtedly, the outcome of these interactions is based on sound biochemical and biophysical principles, a fact well established within the antibiotic literature. In general, three situations can be recognized in which a pair of antimicrobial agents may exert a synergistic effect.

9.3.1   *Inhibition of inactivation*

In this, one compound increases the effective concentration of the other by inhibiting the microbial system responsible for its inactivation. This interaction is most clearly seen in the field of antibiosis where β-lactamase inhibitors are successfully combined with antibiotics containing the β-lactam structure (Jacobs *et al.* 1986). Interactions of such a type are currently unreported in the field of preservation, although resistance through microbial decomposition of preservatives is recorded (Beveridge 1975, Hugo 1987).

9.3.2   *Permeabilization*

A second and more familiar interaction, which forms a common basis for the synergistic effect of many preservatives, is that in which one compound increases the accessibility of targets to another. This usually occurs when one agent exerts its action by increasing the permeability of the cell membrane or cell wall. Besides damaging the cell in its own right, this can permit access of the second compound to targets which were previously concealed. An example of this type of synergy is seen in the co-operative action between benzalkonium chloride, a quarternary ammonium compound which disrupts the cytoplasmic membrane, facilitating access of organo-mercurial agents, such as phenylmercuric acetate, which react with sulphydryl-containing enzymes in the cytoplasm (Hugbo 1977). With this type of synergy it is not necessary for both compounds to possess significant antimicrobial activity. In this context, ethylenediaminetetraacetic acid, having only a weak antibacterial action itself, can enhance the effect of several preservative agents against Gram-negative bacteria, and in particular *Pseudomonas aeruginosa* (Table 9.1), through the chelation of essential bridging cations and consequent destabilization of the

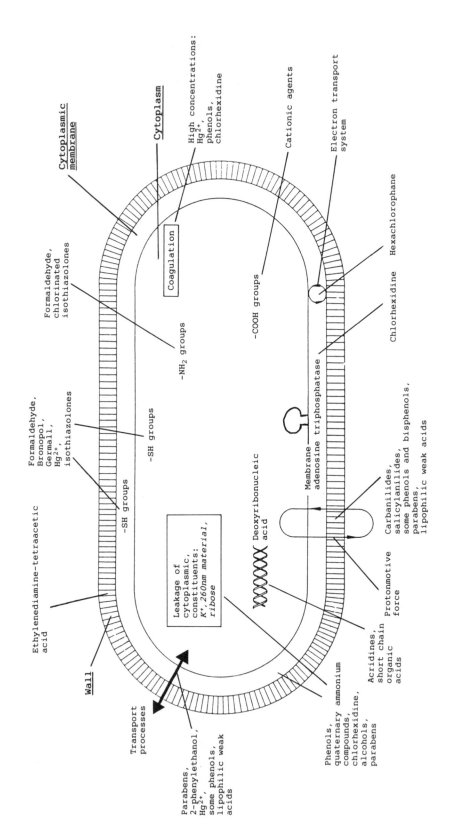

Figure 9.1 Diagram showing major targets for some preservative agents (modified with permission from Hugo and Russell (1992))

Table 9.1   A selection of preservatives whose action is enhanced by ethylenediamine-tetraacetic acid

Benzalkonium chloride
Bronopol
Cetrimide
Chlorhexidine
Chloroxylenol
Germall 115 (imidazolidinyl urea)
Parabens
Phenol
2-Phenylethanol
Sorbic acid

outer membrane barrier (Hart 1984). The role of the cell envelope in resistance to preservative agents is considered in detail in chapter 10.

9.3.3   Biochemical enhancement

The third and potentially broadest category of synergistic interactions is that in which two compounds act simultaneously at different targets, thereby influencing significantly the biochemistry of the cell. Whilst these targets must be different in order to produce synergy they can be quite closely related. For example, the basis of the synergy between chlorocresol and 2-phenylethanol is thought to be through their respective effects on the generation and coupling of a proton gradient to active transport (Denyer *et al.* 1986). A dual action on microbial energetics and chemiosmotic coupling will place severe demands on energy-requiring homeostatic mechanisms and appears a probable basis for synergy (Corner 1981, Moon 1983); it may also play a significant role in preventing the recovery of injured cells. This type of synergy remains the most difficult to predict.

Table 9.2 briefly reports the postulated basis for synergy in several antimicrobial combinations involving preservative agents.

9.4   Measurement of combined action

A variety of methods have been developed to establish the type of interaction occurring between a pair of antimicrobial compounds. While many of these techniques were first applied to, and evaluated with, antibiotic combinations (Beale and Sutherland 1983), they may equally well be used to examine preservative systems (Hodges and Hanlon 1991).

9.4.1   Quantitative assessment

9.4.1.1   Liquid media methods

The most commonly used test method is the *chessboard* (or *chequerboard*) *technique*, so named because of the arrangement of culture tubes employed (Hodges and Hanlon 1991). In this technique, inoculated tubes of nutrient media containing increasing amounts of

Table 9.2   Some studies where a mechanism of synergy has been proposed

Synergistic combinations	Proposed mechanism	Reference
Phenylmercuric acetate and 3-cresol or benzalkonium chloride	Permeabilization	Hugbo (1977)
Chlorhexidine and sulphadiazine	Permeabilization	Quesnel *et al.* (1978)
Chlorhexidine and polymyxin	Biochemical enhancement	Al-Najjar and Quesnel (1979)
Lipophilic weak acids and fatty alcohols	Permeabilization and biochemical enhancement	Corner (1981)
Chlorhexidine and Bronopol	Permeabilization	Wozniak-Parnowska and Krowczynski (1981)
Acetate and lipophilic weak acids	Biochemical enhancement	Moon (1983)
N-chloramines and diazolidinyl urea (Germall 2)	Permeabilization and biochemical enhancement	Llabres and Ahearn (1985)
Chlorocresol and 2-phenylethanol	Biochemical enhancement	Denyer *et al.* (1986)
High and low molecular weight polyhexamethylene biguanides	Specialized permeabilization	Gilbert *et al.* (1990)

antimicrobial agent up to, and slightly above, the *minimum growth inhibitory concentration* (MIC) are arranged such that all possible combinations are tested. After incubation, the resulting pattern of growth can be plotted as an *isobologram* (Fig. 9.2). A straight line between the two MICs indicates a purely additive effect; negative deviation indicates synergy whilst positive deviation shows antagonism. Summing the *fractional inhibitory concentrations* (FIC) of both compounds at the inflection point of the curve gives the ΣFIC—a quantitative measure of the resultant effect. Traditionally, ΣFIC values of less than 0.5 are considered to represent significant synergy but, as pointed out by Berenbaum (1987), this is a purely arbitrary limit. Nevertheless, even minor degrees of synergy (or antagonism) may prove important in marginally preserved products or in cases of a particularly daunting microbial challenge. In performing this technique, particular attention should be paid to the size of the concentration steps between dilutions, since these can exaggerate the resultant effect especially if they are doubling dilutions. Berenbaum (1978) has offered an alternative fixed-ratio dilution technique from which ΣFICs can also be calculated; this has been used in microtitre plate format and correlation with the chessboard technique appears to be generally good (Pons *et al.* 1992). The chessboard technique, when adapted to a microtitre plate system, can be partially automated. It can also be used to measure lethal activity, in which case the *minimum biocidal concentration* and the *fractional biocidal concentration* are used (Quesnel *et al.* 1978).

Bactericidal synergy can also be assessed by determining sterilization times or by following kill curves produced by antimicrobial agents alone and in combination, usually at concentrations achievable under in-use conditions (Richards and Hardie 1972, Akers *et*

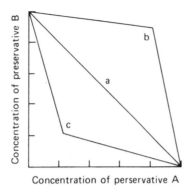

Figure 9.2 Isobologram (growth inhibition) showing three possible results of interaction between preservatives A and B, represented by isobols: curve a, additivity; curve b, antagonism; curve c, synergy

al. 1984, Gilliland *et al.* 1992). Moellering *et al.* (1971), working with antibiotics, defined synergy as a 100-fold increase in killing after 24 h exposure to the combination when compared with the most effective single agent used at the same concentration as in the combination. Again this is a somewhat arbitrary definition, and it may have less relevance in a preservative situation where a rapid killing rate is generally required. In any method involving biocidal action, special attention must be paid to the effective *neutralization* of the antimicrobial agent(s) during the recovery of surviving organisms (Russell 1981).

Drugeon *et al.* (1987) have combined the above techniques to give a measure of the time course of synergy development between two antibiotics.

Turbidometric monitoring of the logarithm phase of growth has been used to follow the effects of antimicrobial agents on bacterial growth both alone and in combination (Brown 1966). The success of the combination is determined from the observed deviation in the expected growth curve or from the increase in mean generation time (Richards and McBride 1971).

9.4.1.2 Solid media method

A quantitative assessment of synergy has been achieved by Schalkowsky and Schalkowsky (1981) using a *spiral plating* device to deposit a continuously decreasing amount of antimicrobial agent onto an agar plate on top of a uniform track of deposited organisms (Hodges and Hanlon 1991). In this way, they have been able to determine inhibitory concentrations for individual compounds and their mixtures.

9.4.2 Qualitative or semi-quantitative assessment

When qualitative rather than quantitative results are adequate, *agar diffusion tests* can be employed. These generally involve two strips or discs of filter paper, each impregnated with a different test antimicrobial compound, placed appropriate distances apart on a seeded agar plate. Following incubation and diffusion, synergy or antagonism can be assessed by the size and shape of the resulting zones of growth inhibition (Maccacaro 1961, Hodges and Hanlon 1991). Hugo and Foster (1963) examined the interactions of 28 combinations of antibacterial agents against six species of organisms using this method. A semiquantitative modification to this basic technique involves a combination of one antimicrobial agent in agar, possibly in the form of a concentration gradient, and the other impregnated in an overlaid paper disc or strip (Hodges and Hanlon 1991). Results of diffusion tests are sometimes difficult to interpret and often depend upon the effects of dilution and ease of diffusion of the compounds in the solid medium.

A further method suitable for studying interactions between preservatives has been described by Wimpenny and co-workers (Wimpenny 1981, Wimpenny *et al.* 1983, Wimpenny and Waters 1984). In this approach the *wedge plate* principle has been extended to generate *two-dimensional diffusion gradient systems* in which the resultant surface microbial growth reflects the pattern of sensitivity to the antimicrobial combination (Wimpenny 1981, McClure and Roberts 1986).

9.4.3 Some comments on the interpretation and use of test data

9.4.3.1 Correlation between test methods and the in-use situation

It is important to recognize when considering *in vitro* test methods that correlation between them is not always obtained since they often measure different parameters and employ different test conditions. Furthermore, extrapolation of the results to the product has to be made with caution since the formulation can further influence the characteristics of a preservative system. To illustrate these points, an example from studies to enhance the preservative effect of a fixed concentration of sorbic acid (0.2%) in an existing acidic cream formulation (Denyer, Hugo and Cavill, unpublished observations) is given in Table 9.3 and Figs 9.3 and 9.4. Here both candidates (chlorhexidine and dehydroacetic acid), showing promise in the agar diffusion approach against *Staphylococcus aureus* (Table 9.3), showed equivalent promise in the chessboard method (Fig. 9.3). Significantly, how-ever, when chlorhexidine and dehydroacetic acid were introduced at bactericidal concen-trations into the formulation, both increased the preservative efficacy as measured by a modification of the test in the *British Pharmacopoeia, Addendum* (Anon. 1982) test but they did so in the reverse order to that expected (Fig. 9.4).

Table 9.3   Combined effect of half the minimum growth inhibitory concentration of sorbic acid and a selection of preservatives against *Staphylococcus aureus* at pH 4 (diffusion assay[*])

Preservative in filter paper strip	Zone of inhibition (mm)	
	No sorbic acid	0.1% sorbic acid
Bronopol (0.1%)	10	12
Chlorhexidine (1%)	9	25
Dehydroacetic acid (6%)	15	25
2-Phenylethanol (7.5%)	0	0

[*] Sorbic acid included in a seeded agar plate overlaid with preservative-impregnated filter paper strips and incubated at 37°C for 24 h before measurement of the zones of inhibition.

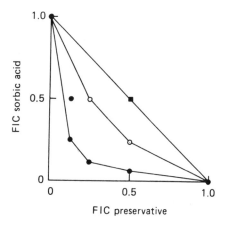

Figure 9.3   Growth inhibitory combinations of various preservatives with sorbic acid against *Staphylococcus aureus* at pH 4: ●, chlorhexidine (MIC, 0.0025%); ○, dehydroacetic acid (MIC, 0.019%); ■, phenylethanol (MIC, 0.45%) or Bronopol (MIC, 0.0031%)

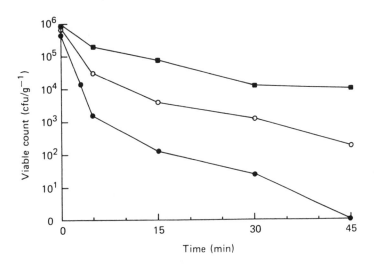

Figure 9.4   Recovery of *Staphylococcus aureus* from a sorbic acid-preserved (0.2% w/v) cream preparation: ■, cream; ○, cream+0.02% chlorhexidine; ●, cream+0.2% dehydroacetic acid

9.4.3.2   *Choice of preservative concentrations within a combination*

Figure 9.2, curve c, illustrates how a specific ratio of preservative agents may be necessary for optimal enhancement of activity. A rather more complex isobologram has been reported (Rehm 1959, Rehm and Stahl 1960) which demonstrates a ratio-dependent alternation between synergy–antagonism and additivity in the same combination (Fig. 9.5). This is not altogether unexpected since many preservatives have concentration-dependent alternative sites of action and this will inevitably affect the basis of synergy (see Fig. 9.1). Thus, it becomes important to establish that synergy (or additivity) is maintained at all concentration ratios; otherwise unequal distribution of preservatives into component

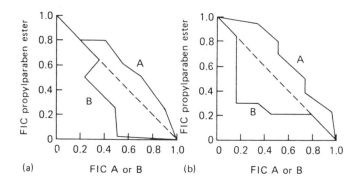

Figure 9.5 Isobolograms of sorbic acid (curves A) and sodium sulphite (curves B) in combination with propyl *p*-hydroxybenzoate ester: (a) *Escherichia coli* (Rehm 1959); (b) *Aspergillus niger* (Rehm and Stahl 1960)

phases of a formulation, for example, may result in an unexpectedly severe reduction in combined activity. Furthermore, since the relationship between antimicrobial activity and concentration for each agent will vary according to its *concentration exponent* (Hugo and Denyer 1987), it is important to establish the efficacy of a combination at its in-use concentration. Similarly, it must be appreciated that dilution of a formulated product in another vehicle, perhaps to provide lower levels of a therapeutic ingredient, may well affect the combined action of preservative agents.

9.4.3.3 *Factorial experimental design*

Clearly, the potential permutations of preservative combination, concentration ratio, product excipient, and formulation factors can be daunting. In preformulation studies, factorial experimental design is to be advocated as part of the screening process to select and optimize preservative combinations (Karabit *et al.* 1989).

9.5 Examples of combined action

Combinations of preservative agents whose interactions have been shown to be synergistic have been reviewed (Denyer *et al.* 1985, Denyer and Wallhaeusser 1990). Representative examples of all types of interaction are given in Table 9.4. In considering these examples, it is important to remember that certain combinations show species specificity (see for example Rehm 1959, Rehm and Stahl 1960, Kull *et al.* 1961, Richards and Hardie 1972, Robaach and Stateler 1980, Toler 1985, Pons *et al.* 1992).

Some advantageous combinations have been exploited commercially, e.g. Phenonip (2-phenoxyethanol and parabens) (Parker *et al.* 1968, Wallhaeusser 1984) and Lauribic (glyceryl monolaurate and sorbic acid) (Kabara 1980, 1984a), and others have been explored in model formulations (Boehm 1968, Rosen *et al.* 1977, Kabara 1984b, Toler 1985). An ongoing survey of preservative systems in commercial ophthalmic preparations confirms the widespread use of benzalkonium chloride/ethylenediamine tetraacetic acid combinations (Anon. 1994). Reference is frequently also made to synergy between different paraben analogues; this appears to be a special case, perhaps determined by individual dose–response relationships (Gilliland *et al.* 1992). Multiple-component

Table 9.4   Some examples of interactions between preservative agents

Preservative	Preservative B		
	Synergistic response	Additive/indifferent response	Antagonistic response
Benzalkonium	Phenylmercuric salt (Hugbo 1976, 1977)	3-cresol (Hugbo 1977, Pons *et al.* 1992)	Hexachlorohane (Walter and Gump 1962)
Benzoic acid	Dehydroacetic acid (Winkler 1955)	Parabens (Rehm and Stahl 1960)	Boric acid (Rehm 1959)
Bronopol	Benzalkonium (Wozniak-Parnowska and Krowzczynski 1981)	Sorbic acid (this chapter)	—
Chlorbutanol	2-phenylethanol (Richards and McBride 1971, 1972)	—	Ethylenediamine tetraacetic acid (Richards and McBride 1972)
Chlorocresol	2-phenylethanol (Denyer *et al.* 1986, Richards and McBride 1971, 1972)	Kathon CG® (Pons *et al.* 1992)	Benzalkonium (Pons *et al.* 1992)
Kathon CG® (methyl isothiazolone chloromethyl isothiazolone)	Chlorhexidine (Pons *et al.* 1992)	2-phenylethanol (Pons *et al.* 1992)	—
Phenylmercuric salts	3-cresol (Hugbo 1977)	—	—
Parabens	Germall (Berke and Rosen 1970, Jacobs *et al.* 1975)	Benzoic acid (Rehm 1959)	Boric acid (Rehm 1959)
Sorbic acid	Dehydroacetic acid (this chapter)	Benzoic acid (Rehm 1959, Rehm and Stahl 1960)	Parabens (Rehm 1959)

preservative systems, with three or more components, have been employed (Wells and Lubowe 1964, Boehm and Maddox 1971, Parker 1973) but their theoretical basis has been questioned (Parker 1982). From a holistic viewpoint the potential for formulation ingredients to enhance the antimicrobial action of a preservative system should be recognized (Kabara 1981, 1984b, Orth 1990, Van Doorne 1990); in this context, the synergistic potential of glycols and phenolic antioxidants is noteworthy.

9.6   Conclusions

Effective preservation of formulated products quite frequently requires the use of combinations of preservative agents. These combinations are not always selected on a rational basis, and thus conditions for preservation are often not optimized. In any attempt to improve the selection of candidates for mixtures, it is important to choose agents with complementary mechanisms of action and physicochemical characteristics. In this way, it should be possible to avoid combinations or concentration ratios which are unfavourable. These 'first-choice' contenders can then be evaluated in the laboratory in experimental systems. It should be remembered, however, that formulation characteristics may alter the availability or efficacy of individual components and promising combinations should always be tested in the final formulation.

References

AKERS, M. J., BOAND, A. V., and BINKLEY, D. A. (1984) Preformulation method for parenteral preservative efficacy evaluation. *J. Pharm. Sci.* **73** 903–905.

AL-NAJJAR, A. R., and QUESNEL, L. B. (1979) Synergism between chlorhexidine and polymyxins against *Pseudomonas aeruginosa. J. Appl. Bacteriol.* **47** 469–474.

ANON. (1982) *British Pharmacopoeia, Addendum.* Efficacy of antimicrobial preservatives in pharmaceutical products. HMSO, London, Appendix XVIC, A22–A24.

(1994) *Monthly index of medical specialities,* July issue. Medical Publications Ltd, London, p. 215.

BEALE, A. S., and SUTHERLAND, R. (1983) Measurement of combined antibiotic action. In: Russell, A. D., and Quesnel, L. B. (eds), *Antibiotics: assessment of antimicrobial activity and resistance* (Society for Applied Bacteriology Technical Series No. 18.) Academic Press, London, pp. 299–315.

BERENBAUM, M. C. (1978) A method for testing for synergy with any number of agents. *J. Infect. Dis.* **137** 122-130.

(1987) Minor synergy and antagonism may be clinically important. *J. Antimicrob. Chemother.* **19** 271–273.

BERKE, P. A., and ROSEN, W. E. (1970) Germall, a new family of antimicrobial preservatives for cosmetics. *Am. Perfum. Cosmet.* **85** 55–60.

BEVERIDGE, E. G. (1975) The microbial spoilage of pharmaceutical products. In: Lovelock, D. W., and Gilbert, R. J. (eds), *Microbial aspects of the deterioration of materials.* Academic Press, London, pp. 213–235.

BLOOMFIELD, S. F. (1986) Control of microbial contamination. Part 2: Current problems in preservation. *Br. J. Pharmacol.* March 72–79.

BOEHM, E. E. (1968) Synergism *in vitro* of certain antimicrobial agents. *J. Soc. Cosmet. Chem.* **19** 531–549.

BOEHM, E. E., and MADDOX, D. N. (1971) Problems of cosmetic preservation. *Manuf. Chem. Aerosol News* **42** 41–43.

BROWN, M. R. W. (1966) Turbidometric method for the rapid evaluation of antimicrobial agents. *J. Soc. Cosmet. Chem.* **17** 185–195.

CORNER, T. R. (1981) Synergism in the inhibition of *Bacillus subtilis* by combinations of lipophilic weak acids and fatty alcohols. *Antimicrob. Agents Chemother.* **19** 1082–1085.

CROSHAW, B. (1977) Preservatives for cosmetics and toiletries. *J. Soc. Cosmet. Chem.* **28** 3–16.

DENYER, S. P. (1990) Mechanisms of action of biocides. *Int. Biodet.* **26** 89–100.

DENYER, S. P., and HUGO, W. B. (eds) (1991) *Mechanisms of action of chemical biocides: their study and exploitation* (SAB Technical Series No. 27). Blackwell Scientific Publications, Oxford.

DENYER, S. P., and WALLHAEUSSER, K.-H. (1990) Antimicrobial preservatives and their properties. In: Denyer, S., and Baird, R. (eds), *Guide to microbiological control in pharmaceuticals.* Ellis Horwood, Chichester, pp. 251–273.

DENYER, S. P., HUGO, W. B., and HARDING, V. D. (1985) Synergy in preservative combinations. *Int. J. Pharmaceut.* **25** 245–253.

— (1986) The biochemical basis of synergy between the antibacterial agents, chlorocresol and 2-phenylethanol. *Int. J. Pharm.* **29** 29-36.

DRUGEON, H. B., CAILLON, J., JUVIN, M. E., and PIRAULT, J. L. (1987) Dynamics of ceftazidine–pefloxacin interaction shown by a new killing curve—chequerboard method. *J. Antimicrob. Chemother.* **19** 197–203.

FAROUK, K. (1981) *In vitro* studies on cinchocaine–preservative combinations. *Int. J. Pharm.* **9** 1–5.

FRANKLIN, T. J., and SNOW, G. A. (1981) *Biochemistry of antimicrobial action*, 3rd edn. Chapman & Hall, London.

GILBERT, P., PEMBERTON, D., and WILKINSON, D. E. (1990) Synergism within polyhexamethylene biguanide biocide formulations. *J. Appl. Bact.* **69** 593–598.

GILLILIAND, D. LI WAN PO, A., and SCOTT, E. (1992) Kinetic evaluation of claimed synergistic paraben combinations using a factional design. *J. Appl. Bact.* **72** 258–261.

HAROLD, F. M. (1970) Antimicrobial agents and membrane function. *Adv. Microb. Physiol.* **4** 45–104.

HART, J. R. (1984) Chelating agents as preservative potentiators. In: Jabara, J. J. (ed.), *Cosmetic and drug preservation: principles and practice* (Cosmetic Science and Technology Series), Vol. 1. Marcel Dekker, New York, pp. 323–337.

HODGES, N. A., and HANLON, G. W. (1991) Detection and measurement of combined biocide action. In: Denyer, S., and Hugo, W. B. (eds), *Mechanisms of action of chemical biocides: their study and exploitation.* (SAB Technical Series No. 27) Blackwell Scientific Publications, Oxford, pp. 297–310.

HUGBO, P. G. (1976) Additivity and synergy *in vitro* as displayed by mixtures of some commonly employed antibacterial preservatives. *Can. J. Pharm. Sci.* **11** 17–20.

— (1977) Additivity and synergism *in vitro* as displayed by mixtures of some commonly employed antibacterial preservatives. *Cosmet. Toilet.* **92** 52, 55, 56.

HUGO, W. B. (1982) Disinfection mechanisms. In: Russell, A. D., Hugo, W. B., and Ayliffe, G A. J. (eds), *Principles and practice of disinfection, preservation and sterilisation.* Blackwell Scientific, Oxford, pp. 158–185.

— (1987) The degradation of preservatives by microorganisms. IN: Houghton, D. R., Smith, R. N., and Eggins, H. O. W. (eds), *Biodeterioration* Vol. 7, Elsevier Applied Science, London, pp. 163–170.

HUGO, W. B., and DENYER, S. P. (1987) The concentration exponent of disinfectants and preservatives (biocides). In: Board, R. G., Allwood, M. C., and Banks, J. G. (eds), *Preservatives in the food, pharmaceutical and environmental industries* (SAB Technical Series No. 22). Blackwell Scientific, Oxford, pp. 281–291.

HUGO, W. B., and FOSTER, J. H. S. (1963) Demonstration of interaction between pairs of antibacterial agents. *J. Pharm. Pharmacol.* **15** 79.

HUGO, W. B., and RUSSELL, A. D. (eds) (1992) *Pharmaceutical microbiology*, 5th edn. Blackwell Scientific, Oxford.

JACOBS, G., HENRY, S. M., and COTTY, V. F. (1975) The influence of pH, emulsifier and accelerated ageing upon preservative requirements of oil/water emulsions. *J. Soc. Cosmet. Chem.* **26** 105–117.

JACOBS, M. R., ARONOFF, S. C., JOHENNING, S., and YAMABE, S. (1986) Comparative activities of the β-lactamase inhibitors YTR 830, clavulanate and sulbactam combined with extended spectrum penicillins against ticarcillin resistant Enterobacteriacae and pseudomonads. *J. Antimicrob. Chemother.* **18** 177–184.

KABARA, J. J. (1980) GRAS antimicrobial agents for cosmetic products. *J. Soc. Cosmet. Chem.* **31** 1–10.

— (1981) Food-grade chemicals for use in designing food preservative systems. *J. Food Protect.* **44** 633–647.

(1984a) Lauricidin. The non-ionic emulsifier with antimicrobial properties. In: Kabara, J. J. (ed.), *Cosmetic and drug preservation. Principles and practice* (Cosmetic Science and Technology Series No. 1). Marcel Dekker, New York, pp. 305–322.

(1984b) Food-grade chemicals in a systems approach to cosmetic preservation. In: Kabara, J. J. (ed.), *Cosmetic and drug preservation. Principles and practice* (Cosmetic Science and Technology Series No. 1). Marcel Dekker, New York, pp. 339–356.

KARABIT, M. S., JUNESKANS, O. T., and LUNDGREN, P. (1989) Factorial designs in the evaluation of preservative efficacy. *Int. J. Pharm.* **56** 169–174.

KULL, F. C., EISMAN, P. C. SYLWESTROWICZ, H. D., and MAYER, R. L. (1961) Mixtures of quaternary ammonium compounds and long-chain fatty acids as antifungal agents. *Appl. Microbiol.* **9** 538–541.

LLABRES, C. M., and AHEARN, D. G. (1985) Antimicrobial activities of N-chloramines and diazolidinyl urea. *Appl. Environ. Microbiol.* **49** 370–373.

MACCACARO, G. A. (1961) The assessment of the interaction between antibacterial drugs. *Prog. Ind. Microbiol.* **3** 173–210.

MCCARTHY, T. J. (1984) Formulated factors affecting the activity of preservatives. In: Karaba, J. J. (ed.), *Cosmetic and drug preservation. Principles and practice* (Cosmetic Science and Technology Series No. 1). Marcel Dekker, New York, pp. 359-388.

MCCLURE, P. J., and ROBERTS, T. A. (1986) Detection of growth of *Escherichia coli* on two-dimensional diffusion gradient plates. *Lett. Appl. Microbiol.* **3** 53–56.

MOELLERING, R. C., Jr., WENNERSTEN, C. B. G., and WEINBERG, A. N. (1971) Synergy of penicillin and gentamicin against enterococci. *J. Infect. Dis.* **124** S207–S209.

MOON, N. J. (1983) Inhibition of the growth of acid tolerant yeasts by acetate, lactate and propionate and their synergistic mixtures. *J. Appl. Bacteriol.* **55** 453–460.

ORTH, D. S. (1990) Principles of preservation. In: Denyer, S., and Baird, R. (eds), *Guide to microbiological control in pharmaceuticals*. Ellis Horwood, Chichester, pp. 241–250.

PARKER, M. S. (1973) Some aspects of the use of preservatives in combination. *Soap, Perfum. Cosmet.* **46** 223–224.

(1982) The preservation of pharmaceutical and cosmetic products. In: Russell, A. D., Hugo, W. B., and Ayliffe, G. A. J. (eds), *Principles and practice of disinfection, preservation and sterilisation*. Blackwell Scientific, Oxford, pp. 287–305.

PARKER, M. S., MCCAFFERTY, M., and MACBRIDE, S. (1968) Phenonip: a broad spectrum preservative. *Soap, Perfum. Cosmet.* **41** 647–649.

PONS, J.-L., BONNAVEIRO, N., CHEVALIER, J., and CRÉMIEUX, A. (1992) Evaluation of antimicrobial interactions between chlorhexidines, quaternary ammonium compounds, preservatives and excipients. *J. Appl. Bact.* **73** 395–400.

QUESNEL, L. B., AL-NAJJAR, A. R., BUDDHAVUDHIKRAI, P. (1978) Synergism between chlorhexidine and sulphadiazine. *J. Appl. Bacteriol.* **45** 397–405.

REHM, H.-J. (1959) Untersuchung zur Wirkung von Konservierungsmittelkombinationen. Die Wirkung einfacher Konservierungsmittelkombinationen auf *Escherichia coli. Z. Lebensm.-Unters. Forsch.* **110** 356–363.

REHM, H.-J., and STAHL, U. (1959) Zur Wirkung antimikrobieller Stoffe in Kombination. *Naturwissenschaften* **46** 431–432.

(1960) Utersuchungen zur Wirkung von Konservierungsmittelkombinationen. Die Wirkung einfacher Konservierungsmittelkombinationen auf *Aspergillus niger* und *Saccharomyces cerevisiae. Z. Lebensm.-Unters. Forsch.* **113** 34–47.

RICHARDS, R. M. E., and HARDIE, M. P. (1972) Effect of polysorbate 80 and phenylethanol on the antibacterial activity of fentichlor. *J. Pharm. Pharmacol.* **24** 90P–93P.

RICHARDS, R. M. E., and MCBRIDE, R. J. (1971) Phenylethanol enhancement of preservatives used in ophthalmic preparations. *J. Pharm. Pharmacol.* **23** 141S–146S.

(1972) The preservation of ophthalmic solutions with antibacterial combinations. *J. Pharm. Pharmacol.* **24** 145–148.

(1973) Preservation of sulphacetamide eye-drops BPC. *Pharm. J.* **210** 118–120.

RICHARDS, R. M. E., and REARY, J. M. (1972) Changes in antibacterial activity of thiomersal and phenylmercuric nitrate on autoclaving with certain adjuvants. *J. Pharm. Pharmacol.* **24** 84P–89P.

ROBAACH, M. C., and STATELER, C. L. (1980) Inhibition of *Staphylococcus aureus* by potassium sorbate in combination with sodium chloride, tertiary butylhydroquinone, butylated hydroxyanisole or ethylenediamine tetraacetic acid. *J. Food Protect.* **43** 208–211.

ROSEN, W. E., BERKE, P. A., MATZIN, T., and PETERSON, A. F. (1977) Preservation of cosmetic lotions with imidazolidinyl urea plus parabens. *J. Soc. Cosmet. Chem.* **28** 83–87.

RUSSELL, A. D. (1981) Neutralisation procedures in the evaluation of bactericidal activity. In: Collins, C. H., Allwood, M. C., Bloomfield, S. F., and Fox, A. (eds), *Disinfectants. Their use and evaluation of effectiveness* (SAB Technical Series No. 16). Academic Press, pp. 45–59.

SCHALKOWSKY, S., and SCHALKOWSKY, E. (1981) Application of the spiral plating method to bacterial interaction tests. *Third International Symposium on Rapid Methods and Automation in Microbiology, May 27, 1981, Washington, DC.*

TOLER, J. C. (1985) Preservative stability and preservative systems. *Int. J. Cosmet. Sci.* **7** 157–164.

VAN DOORNE, H. (1990) Interactions between preservatives and pharmaceutical components. In: Denyer, S., and Baird, R. (eds), *Guide to microbiological control in pharmaceuticals.* Ellis Horwood, Chichester, pp. 274–291.

WALLHAEUSSER K. H. (1984) Antimicrobial preservatives used by the cosmetic industry. In: Kabara, J. J. (ed.), *Cosmetic and drug preservation. Principles and Practice* (Cosmetic Science and Technology Series No. 1). Marcel Dekker, New York, 1984, pp. 605–745.

WALTER, G. R., and GUMP, W. S. (1962) Antibacterial activity of mixtures of quaternary ammonium compounds and hexachlorophene. *J. Pharm. Sci.* **51** 770–772.

WELLS, F. V., and LUBOWE, I. I. (1964) In: *Cosmetics and the skin.* Reinhold, New York. pp. 586–598.

WIMPENNY, J. W. T. (1981) Spatial order in microbial ecosystems. *Biol. Rev.* **56** 295–342.

WIMPENNY, J. W. T., and WATERS, P. (1984) Growth of micro-organisms in gel-stabilized two-dimensional diffusion gradient systems. *J. Gen. Microbiol.* **130** 2921–2926.

WIMPENNY, J. W. T., LOVITT, R. W., and COOMBS, J. P. (1983) Laboratory model systems for the investigation of spatially organised microbial ecosystems. *Symp. Soc. Gen. Microbiol.* **34** 67–117.

WINKLER, J. (1955) *Method and composition for inhibiting the growth of micro-organisms*, US Patent 2.722,483.

WOZNIAK-PARNOWSKA, W., and KROWCZYNSKI, L. (1981) New approach to preserving eye drops. *Pharma. Int.* **2** 91–94.

10

Microbial resistance to preservative systems

PETER GILBERT AND JULIE A. DAS

Department of Pharmacy, University of Manchester, Oxford Road, Manchester M13 9PL

Summary

(1) Resistance may be innate and reflect fundamental differences in structure rendering one group of organisms sensitive to particular agents, or it might be acquired and reflect modification of target sites and/or barriers to drug permeation external to those sites such as the cell envelope.

(2) The physiological status of the bacterial cells and the nature of the ecosystems in which they have grown can lead to marked changes in their sensitivity to chemical antimicrobial agents.

(3) The effects of cellular growth-rate, the nature of the growth-limiting nutrient, the presence of extracellular polysaccharides and growth as biofilms are primary modulators of preservative sensitivity.

10.1 Introduction

10.1.1 *Preservative failure*

Failure of a preservative agent in practice might be attributed to a number of causes.

Firstly the agent might be present within the aqueous (biological) phase of the formulation at an insufficient concentration to prevent microbial growth. This might be related to interaction of the preservative with excipients of the formulation (e.g. container, surfactants, colloidal materials, and salts) or, alternatively, the pH of the product might, for an ionizable preservative, cause an inactive species (i.e. unionized preservative) to predominate or the hydrophobic character of the preservative might cause it to accumulate within the non-biological oil-phase of the formulation. In the first instance, neutralization or loss of preservative can be slow and allow the eventual out-growth of persistent survivors or render the product susceptible to extrinsic contamination. The outgrowing survivors need not necessarily have acquired especial resistance characteristics to the preservative in question but would generally have some degradative ability with respect to the formulation. In this latter respect, destabilization and biodeterioration of the formulation might render it susceptible as a growth medium to organisms incapable of utilizing the original formulation.

Secondly, preservative failure might result from the presence of micro-organisms within the formulation which are resistant to its antimicrobial effect. Resistance of such micro-organisms might arise as a result of phenotypic changes in the cells; for example, cell susceptibility can be significantly altered as a result of changes in gross cellular composition as a response to the growth environment. Slowly growing cells generally survive antimicrobial treatments better than those which are replicating more rapidly (Brown *et al.* 1990). The effects of decreased growth rates are often coupled to the availability of essential nutrients and also as to whether the organisms are growing attached to a surface or free from surfaces. Microbial resistance, resulting from these varied criteria is the subject of this chapter, and each of these factors will be expanded in detail.

10.1.2   *Microbial resistance*

In order to be active, all antimicrobial agents must be capable of reaching their bio-chemical targets. These are commonly located within the cell envelope and cytosol rather than being on the immediate outer surface of the cells. All agents must therefore interact with the outer layers of the cell, permeating them, at least in part, in order to concentrate at their targets. Resistance to the action of antimicrobial agents therefore often depends upon the permeability of the cell envelope towards antimicrobial agents. In this respect, resistance can be innate and represent fundamental differences in envelope composition between the various groups of organisms, such as Gram-positive and Gram-negative bacteria, or it might be acquired and reflect changes in the envelope in response to environmental stimuli. In the latter instance, these changes might alter the abundance of a target within the envelope, rendering cells resistant, or it might affect drug permeation through it. In addition, it is now generally recognized that bacteria commonly grow in association with surfaces as adherent biofilms (Costerton *et al.* 1987). Biofilms consist of functional consortia of cells, often expressing unique surface-associated physiologies, and organized within extensive exopolymer matrices (glycocalyx). The glycocalyx will not only provide a further permeability barrier towards preservatives but will also concentrate various extracellular products of the cells which might influence preservative activity (Brown and Williams 1985, Gilbert *et al.* 1990).

It is the intention of this chapter to examine the passage of chemical preservatives across the glycocalyx and cell envelope to the cytoplasmic membrane and cytosol and to discuss the extent to which phenotypic modification of the cell can affect this passage and thereby influence resistance towards preservatives of pharmaceutical and cosmetic products.

10.2   Modulation of preservative effectiveness by the cell envelope and glycocalyx

Since they are often free-living, bacteria must interact with an ever-changing and often hostile environment. Their immediate contact with the environment is through the cell envelope. The effect of such changes to the cells are minimized through compensatory changes in their cell envelopes (i.e. cytoplasmic membrane, cell wall and if present, glycocalyx (Fig. 10.1)).

10.2.1 Cell envelope

The cell envelope, an integral part of the cell, modulates access of nutrients and to some extent preservative compounds to the underlying cytoplasmic membrane. Two distinct mechanisms are thought to be involved in such action:

(1)   The cytoplasmic membrane is bathed in a fluid continuous with that of the surrounding medium. The peptidoglycan matrix in Gram-positive bacteria (Scherrer and Gerhardt 1971) and in conjunction with the outer membrane in Gram-negative ones (Nikaido 1976, Nikaido and Nakae 1979) intercedes and functions as a molecular sieve, thereby preventing the ready access of large hydrophilic molecules to the cell membrane.

(2)   Where such ready access to the membrane is denied, then a drug must interact with the cell wall in order to traverse it. With Gram-negative bacteria, this will involve sequential partitioning across the lipophilic layers that constitute the outer-membrane.

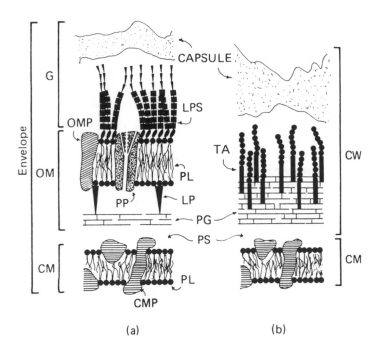

(a)   (b)

Figure 10.1   Diagrammatic representation of the (a) Gram-negative and (b) Gram-positive cell envelopes: CW, cell wall; PS, periplasm; CM, cytoplasmic membrane; CMP, cytoplasmic membrane protein; OM, outer-membrane; G, glycocalyx; OMP, outer-membrane protein; LPS, lipopolysaccharide; PL, phospholipid; LP, lipoprotein, PP, porin protein; TA, teichoic acid; PG, peptidoglycan

Fundamental structural differences exist between Gram-negative and Gram-positive cell walls. The outer-membrane of Gram-negative cells renders them impermeable to many hydrophilic agents relative to Gram-positive cells. Such fundamental differences, are often referred to as intrinsic or innate resistance and are illustrated Table 10.1. In view of these major differences it is necessary to consider separately the contributions of Gram-negative and Gram-positive cell envelopes towards preservative resistance. A general consequence of these differences, however, is that the organisms encountered in pharmaceuticals, cosmetics and toiletries which demonstrate resistance towards commonly employed preservatives, such as the phenolics, quaternary ammonium compounds and *p*-hydroxybenzoates, are often Gram-negative in nature.

10.2.2 Modulation of preservative effectiveness by the Gram-positive cell wall

The Gram-positive cell wall (Fig. 10.1) is composed of a complex, highly cross-linked network of peptidoglycan. Incorporated into this matrix are the teichoic and teichuronic acids (Hess and Lagg 1958, Gerhardt and Judge 1964). The polysaccharide components of the teichoic and teichuronic acids radiate into the surrounding medium and contribute to the net negative charge of the bacterial surface and hinder the access of hydrophobic agents to the cytoplasmic membrane. Additionally, the teichoic and teichuronic acid residues are thought to be associated with the initial binding, prior to uptake, of cationic nutrients and might also affect adsorption of cationic antimicrobial agents to the cell surface. In this light, Meers and Tempest (1970) and Tempest *et al.* (1968) showed that magnesium limitation in the growth environment caused an increased affinity of the cell wall for magnesium and that this was associated with changes in teichoic acid and teichuronic acid content.

The function of the peptidoglycan matrix as a molecular sieve depends greatly upon thickness and the extent of cross-linking (Mitchell 1959, Scherrer and Gerhardt 1971). Both of these properties are dependent upon the physiological status of the cells, particularly growth rate and the nature of the growth-limiting nutrient. Cross-linking and *de novo* peptidoglycan biosynthesis occur independently in the cell. In fast-growing cells, peptidoglycan biosynthesis proceeds at a faster rate than cross-linkage, whilst for cells in stationary phase cross-linking proceeds with little *de novo* synthesis. The degree of cross-linkage and thereby porosity of the peptidoglycan therefore show inverse and direct

Table 10.1 Relative intrinsic resistance of Gram-positive and Gram-negative bacteria towards some commonly used chemical preservatives

Agent	Minimum growth inhibitory concentrations (µg/ml)	
	Staphylococcus aureus	*Escherichia coli*
Chlorhexidine diacetate	0.3–1.0	1.0–1.2
Dequalium chloride	0.6	8.0
8-Hydroxyquinolone	2.5–4	6.2–6.4
Cetrimide	0.39–4.0	12.5–16
Hexachlorophane	0.05–0.5	12.5–100.0

Data taken from Lambert and Hammond (1983) and Wallhaeusser (1984).

relationships to growth rate, respectively. The combined effects of the degree of cross-linkage, produced through change of growth rate, and the effects of nutrient-limitation are illustrated in Fig. 10.2 for the action of the preservatives, chlorhexidine and 2-phenoxyethanol upon vegetative *Bacillus megaterium* cells. Sensitivity data determined for the protoplasts of similarly grown cells (Fig. 10.2), illustrate that the change in preservative sensitivity can be attributed, in this instance, to alteration of the cell wall.

The size exclusion limit of peptidoglycan is generally estimated to be in the order of 100,000 daltons. This can be significantly decreased, however, through the presence of charged species in the peptidoglycan matrix (Marquis 1968). Thus, Nadir and Gilbert (1980) observed that the activity of chlorhexidine could be significantly decreased through the presence of potassium ions and other cations and that this reduction in activity was associated with decreased adsorption of the agent by the cells. Such work suggests that, in the presence of salts (0.15 M), contraction of the matrix and competition for binding had significantly reduced permeation by the biocide.

10.2.3 Modulation of preservative effectiveness by the Gram-negative cell wall

There have been a number of excellent reviews detailing the molecular architecture of the Gram-negative cell envelope and the passage of drugs across it (Brown *et al.* 1990, Costerton 1977, Nikaido and Nakae 1979, Lugtenberg and Van Alphen 1983). The Gram-negative cell envelope (Fig. 10.1) consists essentially of an asymmetric membrane bilayer, comprising lipopolysaccharide and phospholipid, attached via Braun's

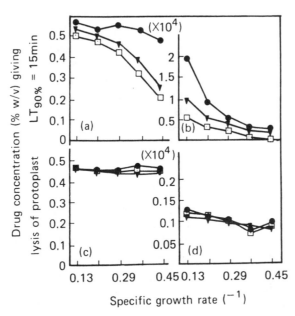

Figure 10.2 Influence of nutrient limitation and growth rate upon the sensitivity of (a) and (b) *Bacillus megaterium* cell, (c) and (d) sphaeroplast suspensions towards bactericidal concentrations of (a) and (c) 2-phenoxyethanol, (b) and (d) chlorhexidine: ●, phosphate-limited cultures; □, magnesium-limited cultures; ▼, carbon-limited cultures. (Reprinted, with permission of Blackwell Scientific Publications, from Gilbert and Brown (1980))

lipoproteins and porin proteins to a thin matrix of peptidoglycan (Nikaido and Nakae 1979, Osborn and Wu 1980). The outer leaflet of this outer-membrane is predominately close-packed lipopolysaccharide orientated such that the hydrophilic polysaccharide chains radiate into the surrounding medium. The lipophilic portions are directed inwards and interact with the fatty-acid regions of a phospholipid monolayer, which constitutes the inner-leaflet of the outer membrane. The two leaflets form a central hydrophobic lipid core to the outer membrane. Fluidity within this hydrophobic core is very low and attributed to the close packing of the lipopolysaccharide: lipid A components and stacking of the ketodeoxyoctonic acid residues (Nikaido and Nakae 1979), which severely restricts dissolution and diffusion of hydrophobic drug molecules into and through it.

Stability of the outer membrane–peptidoglycan complex is maintained by the presence of lipoproteins and divalent cation bridges (De Pamphilus 1971, Schnaitman 1971a, b, Costerton *et al.* 1974). The polysaccharide chains of the lipopolysaccharide attract various cations such as Mg^{2+}, Ca^{2+}, and Zn^{2+}. Concentration of cations acts as an initial step in their uptake and assimilation by the cell and also stabilizes natural electrostatic repulsions which exist within the polysaccharide matrix. The latter effect is through cation-bridging across negatively charged groups on adjacent polysaccharide chains (Nikaido and Nakae 1979). In *Pseudomonas aeruginosa*, there is evidence that the divalent cation bridges play an increased role in stability of the outer membrane complex (Meadow 1975), hence the potential of ion-chelators such as ethylenediaminetetraacetic acid, which remove divalent cations such as magnesium from the envelope, as synergists in preservative combinations. The action of ethylenediaminetetraacetic acid destabilizes the outer membrane and allows access of agents, otherwise excluded from the cell, to the cytoplasmic membrane.

The stabilized polysaccharide chains of the lipopolysaccharide form a tightly clustered, highly hydrophilic covering to the cell which hinders the approach of hydrophobic drug molecules to the outer membrane core. This property is demonstrated by increased sensitivity of Gram-negative organisms towards hydrophobic antibiotics, such as novobiocin, rifampicin and erythromycin, when the polysaccharide chain length of the lipopolysaccharide is reduced through mutation (Tamaki *et al.* 1971, Lieve 1974, Roantree *et al.* 1977). Since the activity of hydropholic agents is unaffected by such mutational change, then the data suggests that separate pathways exist for the diffusion of hydrophilic and hydrophobic agents across the cell envelope.

10.2.3.1 *Diffusion of hydrophilic agents*

The lipophilic core of the outer membrane constitutes a formidable barrier to the penetration of hydrophilic agents. These are able to traverse it, however, through the presence of a number of integral proteins inserted into the outer membrane complex. Outer-membrane-protein-modulated diffusion of hydrophilic agents occurs by two distinct mechanisms. Some material may be transported by specific, transport proteins analogous to those of the cytoplasmic membrane (Sabet and Schnaitman 1973, Hantke 1976, Hancock 1984, Wookey and Rosenberg 1978), whilst others permeate the outer membrane in a less specific manner by passive diffusion through porin proteins (Payne and Gilvarg 1968, Decad and Nikaido 1976, Nikaido and Nakae 1979). An average Gram-negative cell contains approximately 10^5 molecules of porin protein (Rosenbusch 1974), of which several types may be represented (reviewed by Benz 1985, Nakae 1986, Nikaido and Vaara 1985). These proteins are aggregated in the outer membrane as trimers, each of which forms a hydrophilic water-filled channel allowing the non-specific passage of small hydrophilic molecules across the lipid core of the outer membrane (Yu *et al.* 1979, Nakae *et al.* 1979). Rates of transpore diffusion depend upon molecular size (Nikaido and Nakae 1979) and charge (Hancock *et al.* 1986, Benz *et al.* 1985) of the molecule. For molecules approaching the maximum size limit of the channels (a molecular weight of about 600) (Zimmerman and Rosselet 1977), then lipophilicity plays a determining role in the rates of passage. The extent and nature of porin protein distribution in the cell envelope vary significantly as a function of the growth rate and the physiological status of the cells (Bassford *et al.* 1977, Turnowsky *et al.* 1983). Evidence suggests that at any one time a proportion of the porin channels might be closed; thus the cell is able to regulate, in a limited fashion, the extent of transpore diffusion (Benz and Hancock 1981).

10.2.3.2   Diffusion of hydrophobic agents

Biological membranes generally allow the passive diffusion of small hydrophobic molecules across them. The outer membrane is exceptional in this respect in that the hydrophilic covering provided by the lipopolysaccharide partly excludes such molecules from interacting with the membrane surface. The outer membrane of deep rough strains are deficient in the polysaccharide component of the lipopolysaccharide and resemble much more closely typical biological membranes. Hydrophobic molecules may pass across these through a combination of oil–water partitioning and passive diffusion. Hansch and Fujita (1964), Lien *et al.* (1968) and Hansch (1973) argued that, for chemically related drugs active at similar targets, activity was determined by the relative ease of entry into the cell. If the envelope components of the cell are considered as a series of alternating hydrophilic and hydrophobic compartments, then substances of low water solubility would be unable to penetrate the aqueous layers and substances of low lipid solubility would be unable to cross lipid regions. Hansch therefore argued that compounds between these two extremes must exist which possessed the optimal balance of hydrophilic and hydrophobic character; this would allow maximal permeation of the envelope compartments. He went on to demonstrate in a classic series of papers that, if the lipophilicity (log partition coefficient, i.e. log P) of a series of related drugs was plotted against biological activity, then a parabolic relationship often existed, where log P_0 was the value of log P for the compound with maximal activity. In this fashion, it was possible to determine the optimal lipophilicity with which drugs could permeate the cell envelope and reach their targets. Average log P_0 values have been found to be approximately 4.0 for Gram-positive cells and approximately 6.0 for Gram-negative ones (Lien *et al.* 1968). This difference can be attributed to the reduced amounts of lipid present within the Gram-positive cell wall. Clearly therefore changes in the fluidity and lipophilicity of any envelope compartment must affect log P_0 and thereby the activity of single antimicrobial agents.

The bacterial cell is remarkably flexible in both structure and composition and constantly interacts with its environment. Whilst this confers a great survival advantage upon the cells in an ever-changing ecosystem, it will also significantly influence the susceptibility of the cells to chemical inactivation through changes in the polysaccharide chain length of the lipopolysaccharide, changes in the relative amounts of polar and non-polar lipids in the outer membrane or in the cytoplasmic membrane, changes in the degree of saturation of these lipids and fatty-acid chain length and in porin protein composition.

10.2.4 *Modulation of preservative sensitivity through modification of the cell envelope*

Given optimal growth conditions, organisms such as *Escherichia coli* and *Pseudomonas aeruginosa* are able to grow rapidly and efficiently with generation times as short as 20 min. Such rapid growth and division are unlikely to persist in a natural environment for any significant length of time, if indeed they occur at all. More probably the rate and extent of growth are governed by the availability of critical nutrients. Under nutrient-limited conditions, the physiology of the cells adapt in a number of ways.

(1)   rationalization of the usage of the growth-limiting nutrient through using alternative substrates, modification of cell composition and/or reduction in the amounts of macromolecules containing these substances;

(2)   alteration of the cell surface to increase the affinity for the growth-limiting material, thus making uptake into the cytosol more efficient and competitive;

(3)   reduction of the cellular growth rate to the maximum permissible (given (1) and (2)).

The effect of different growth-limiting nutrients therefore is to give rise to cells with different growth rates and coincidentally radically different cell envelopes (Holme 1972, Ellwood and Tempest 1972, Brown and Melling 1969a,b, Brown 1975). This greatly influences susceptibility to antimicrobial agents (Brown 1977, Gilbert and Wright 1986, Brown *et al.* 1990, Gilbert *et al.* 1990, Brown and Gilbert 1993) and antibiotics (De La Rosa *et al.* 1982, Brown and Williams 1985,Toumanen *et al.* 1986) and has been widely reported for a range of bacterial species (Tempest *et al.* 1968; Meers and Tempest 1970, Minnikin *et al.* 1971, Dean 1972, Dean *et al.* 1976, Holme 1972, Broxton *et al.* 1984). With Gram-negative bacteria such changes involve particularly modification of the outer and cytoplasmic membranes (Ellwood and Tempest 1972, Nikaido and Nakae 1979; Lugtenberg and Van Alphen 1983). In this respect, resistance patterns have been associated with altered phospholipid content (Pechey *et al.* 1974, Teuber and Bader 1976, Imai *et al.* 1975, Ikeda *et al.* 1984), porin protein composition (Harder *et al.* 1981, Williams *et al.* 1984), lipopolysaccharide content (Tamaki *et al.* 1971) and cation contents (Brown and Melling 1969a, b, Gilleland *et al.* 1974, Melling *et al.* 1974, Nichols *et al.* 1989, Boggis *et al.* 1979, Shand *et al.* 1985) of the cell envelopes. These are thought to modify the action of chemical antimicrobial agents through the following:

(1)   Changes in the relative abundance of the target molecule/material.

(2)   Alteration of surface charge and thereby initial drug adsorption.

(3)   Alteration of the fluidity and lipophilicity of the various envelope compartments, thereby affecting optimal drug lipophilicity (log P_0).

(4)   Alteration of porin-protein expression and content, thereby affecting permeation of small hydrophilic agents.

The nutritional status of cells growing within their natural habitats is almost impossible to assess (Brown 1977, Gilbert *et al.* 1987, Brown *et al.* 1990). For contaminants found in pharmaceuticals, cosmetics and toiletries, the original growth conditions of the organisms may include the manufacturing water, raw materials, product residues within the manufacturing plant and other parts of the manufacturing environment. One can only speculate about the nature of particular nutrient limitations associated with particular localized sites. In the laboratory such nutrient limitations can be created in both batch and continuous culture through the use of chemically defined media in which all the nutrients except one are present to excess and the concentration of the remaining nutrient is calculated to force the culture into its stationary phase whilst oxygen is still in excess. This latter nutrient becomes the growth-limiting nutrient and has in many studies been chosen as a carbon, nitrogen, magnesium, iron or phosphorus source.

10.2.4.1 Batch culture studies

There have been a number of studies which set out to demonstrate the interrelation of the physiological status of microbial cells, envelope composition and susceptibility towards chemical preservatives. Growth of Gram-negative species under conditions of phosphorus-limitation results in decreases in cellular phospholipid content, compensated for by increases in fatty and neutral lipid content (Gilbert and Brown 1978a). The reduction in phospholipid is generally attributed to loss of diphosphatidylglycerol and phosphatidylethanolamine. Under magnesium-limited conditions, phospholipid content is slightly increased and attributed to increases in diphosphatidylglycerol (Gilbert and Brown 1978a,b). Broxton *et al.* (1984) and Ikeda *et al.* (1984) showed that the polymeric biguanide, Vantocil, interacts specifically with acidic phospholipids of the cytoplasmic membrane. Changes in the relative proportions of the neutral and phosphatide lipids associated with growth under magnesium-limited and phosphorus-limited conditions therefore markedly affected sensitivity towards this and related compounds. Similar mechanisms have been implicated in resistance towards gentamicin and polymyxin (Pechey *et al.* 1974, Finch and Brown 1975, Wright and Gilbert 1987c), the initial binding and action of which are mediated through membrane phospholipids. Both these mechanisms involve changes in the extent of expression of a previously identified target. Other workers have been unable to attribute resistance to particular envelope components. Thus, Gilbert and Brown (1978a) showed that, in batch culture under carbon-limited conditions, *E. coli* were particularly sensitive to the actions of substituted phenols and 2-phenoxyethanol. Klemperer *et al.* (1980) showed that glucose limitation led to increased sensitivity towards chlorhexidine over that of phosphate- and magnesium-limited cultures. The latter data was similar to that observed for the related biguanide, Vantocil, and might also be explained in terms of altered acidic phospholipid content (Broxton *et al.* 1984). Similar results to these were obtained for the organisms *Proteus mirabilis* where magnesium and phosphorus limitation of growth increased resistance towards cetrimide and phenol relative to carbon-limited cultures (Klemperer *et al.* 1977).

Al-Hiti and Gilbert (1980) examined the response of the antimicrobial agents effectiveness test for micro-organisms in the *United States Pharmacopoeia* to a number of commonly used preservatives, after growth in liquid media producing either carbon-, nitrogen-, phosphorus- or magnesium-limited conditions. The results are summarized in Table 10.2. With the exception of *Aspergillus niger*, the preservative sensitivity of the test strains altered significantly with the nature of the nutrient limitation. Least variation was observed with the thiol interactive agent thiomersal and the greatest with benzalkonium chloride. Notably, the organism *P. aeruginosa* was the most resistant to all the agents included in the study. Iso-effective concentrations, towards benzalkonium chloride, varied for this organism from 2.5×10^{-3} (%w/v) for citrate-grown, phosphorus-limited cells to 5×10^{-2} (%w/v) for glycerol-grown, carbon-limited cells. Variation in preservative sensitivity for the remaining organisms, except *Staphylococcus aureus*, was

Table 10.2 Effect of various nutrient limitations in batch culture upon the ability of the organisms to survive and grow in the presence of various concentrations of preservatives

Nutrient-limitation:	Preservative concentration (%w/v $\times$ 10^{-4}) required to reduce the number of colony forming units by 90%								
	Thiomersal			Chlorhexidine			Benzalkonium chloride		
	Nitrogen	Carbon	Phosphorus	Nitrogen	Carbon	Phosphorus	Nitrogen	Carbon	Phosphorus
Organism									
Escherichia coli	0.53	0.37	0.60	24.0	13.0	20.0	100	65	82
Pseudomonas aeruginosa									
(Citrate)	1.90	1.40	0.25	28.0	24.0	30.0	100	450	25
(Glycerol)	5.00	5.00	3.70	62.0	45.0	41.0	500	500	180
Staph. aureus	0.12	0.54	0.14	8.0	10.5	10.5	0.5	0.5	0.6
Candida albicans	0.009	0.009	0.013	10.5	11.0	10.5	52	52	35
Aspergillus niger	0.025	0.025	0.025	2.3	2.3	2.3	2.5	2.5	2.5

From Ali-Hiti and Gilbert (1980)

significant but never exceeded three times the minimum effective concentration. No universal pattern of nutrient limitation and preservative sensitivity was observed between species, but general patterns emerged for the three preservatives tested against each organism. Thus the sensitivity was as follows; for *E. coli*, carbon-limitation > nitrogen-limitation > phosphorus-limitation; for *P. aeruginosa*, phosphorus-limitation > nitrogen-limitation > carbon-limitation; and for *S. aureus*, nitrogen-limitation > phosphorus-limitation > carbon-limitation.

Whilst the use of batch cultures in these studies prevented correlations between envelope composition and preservative sensitivity to be drawn, the inescapable conclusion to be made is that the sensitivity of typical nutrient broth grown cells bears little resemblance to that of cells found in natural habitats where the availability of nutrients restricts growth of the cells (Gilbert *et al.* 1987, Brown *et al.* 1990, Gilbert *et al.* 1990).

10.2.4.2 *Continuous culture studies*

When provided with an excess of all critical nutrients, cells will grow at their maximum growth rate. This can be as fast as three generations per hour and is easily demonstrated in batch culture using nutritionally rich growth media. A major consequence of deprivation of nutrients such as phosphorus, magnesium, carbon or nitrogen sources is a reduction of the cellular growth rate. Growth rate may also be affected by growth in sub-optimal pH, temperatures or water activities. Reduced growth rate enables the cells to contain lower quantities of RNA, ribosomes, anabolic and catabolic enzymes. Thus, where nutrients are required as cofactors for growth as well as being components of new cellular material, then economies will be made upon the use of particular nutrients (e.g. magnesium stabilizes ribosomes and also is a major structural component of the cell envelope). The nature and extent of specific transport proteins associated with non-limiting nutrients will also be rationalized in the light of nutrient deprivation. The growth rates of cells *in situ* and *in vivo* is difficult to assess but, where estimates have been made, then these are only rarely equivalent to those achieved in batch culture and are commonly equivalent to generation times of the order of 20–30 hours (Eudy and Burroughs 1973, Weinberg 1984). There are a number of ways in which growth rate can be controlled in the laboratory. Alteration of medium pH or incubation temperature to sub-optimal values or decreases in the 'richness' of the medium will dramatically reduce growth-rate. In these situations, however, the altered physiology of the cells is more likely to represent adaptation to the altered physical environment than to be a consequence of growth rate *per se*. Thus alteration of the growth temperature will affect the degree of saturation of the fatty acid components of the cytoplasmic and outer membranes and changes in the carbon source will affect the lipid composition of the cells (Hugo and Franklin 1968, Hugo and Davidson 1973, Bullmand and Stretton 1975). These changes have been re-ported to alter the degree of permeation and disposition of preservatives throughout these cells and affect their sensitivity dramatically.

The chemostat (Herbert 1956) allows specific growth rate (μ) to be controlled without any associated changes in the physicochemical environment of the cells. Various workers have used such systems to evaluate the effects of growth rate upon the sensitivity of cells towards antibiotics, disinfectants and preservatives. A general conclusion to be drawn from such studies is that slowly growing cells are particular recalcitrant to chemical inactivation (Brown 1975, Finch and Brown 1975, Gilbert and Brown 1978b, 1980, Brown *et al.* 1990, Gilbert *et al.* 1990).

Finch and Brown (1975) observed that *P. aeruginosa* became particularly sensitive to polymyxin and ethylenediaminetetraacetic acid as the growth rate was increased. Gilbert and Brown (1978b) assessed the sensitivity of this organism towards various substituted phenols and observed significant alteration in sensitivity with changing growth rate (Fig. 10.3). In this study the change in sensitivity was associated with marked alteration in the cellular lipolysaccharide content, as indicated by ketodeoxyoctonic acid content. Increases in lipopolysaccharide resulted in decreased drug uptake by the cells and decreased sensitivity (Fig. 10.4). It was suggested by these workers that the lipopolysaccharide layer formed a barrier in the outer membrane against adsorption of these agents.

Wright and Gilbert (1987a) investigated the interrelation of the chlorhexidine sensitivity of *E. coli* and the growth rate for four nutrient-limitations. Nitrogen- and carbon-limited cultures showed an overall increase in sensitivity as growth rate increased whilst magnesium- and phosphorus-limited cultures showed an opposite trend of increased resistance. At the extremes of growth rate tested, different orders of sensitivity were observed between nutrient limitations. When $\mu < 0.08 \text{ h}^{-1}$, sensitivity was seen to decrease with different nutrient limitations in the sequence carbon-limitation > phosphorus-limitation > magnesium-limitation > nitrogen-limitation, whilst at faster growth rates ($\mu > 0.4 \text{ h}^{-1}$), the sequence was altered to carbon-limitation > nitrogen-limitation > phosphorus-limitation > magnesium-limitation. Overall, carbon-limited cultures were most sensitive towards chlorhexidine, with this limitation showing the least dependency upon growth rate (Fig. 10.5). If chlorhexidine binding and activity were dependent upon acidic phospholipid content or some other cell envelope component, as had been suggested by some earlier studies, then these ought to have demonstrated opposite dependencies upon growth rate for magnesium- and phosphorus-limited cultures than for carbon- and nitrogen-limited cultures. No such correlation was observed either for specific phospholipids (Fig. 10.6), lipopolysaccharide outer-membrane protein composition (Fig. 10.7) or for ratio of phospholipids to fatty and neutral lipids (Fig. 10.8). All these properties changed significantly, however, with growth rate and nutrient

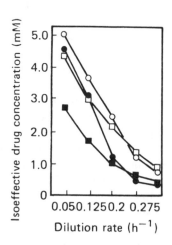

Figure 10.3 Effect of nutrient limitation and growth rate upon the sensitivity of *Pseudomonas aeruginosa* to uncoupling concentrations of 3-chlorophenol (o, •) and 4-chlorophenol (□, ■): o, □, magnesium-limited cultures; •, ■, glucose-limited cultures. (Reprinted, with permission of The American Society for Microbiology, from Gilbert and Brown (1978b))

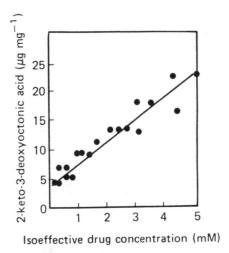

Figure 10.4 Interrelationship of the sensitivity of *Pseudomonas aeruginosa* suspensions to chlorinated phenols and their lipopolysaccharide content, as estimated as 2-ketodeoxyoctonic acid content. (Reprinted, with permission of The American Society of Microbiology, from Gilbert and Brown (1978b))

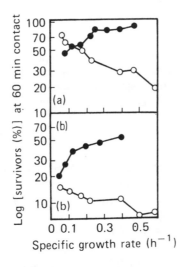

Figure 10.5 Bactericidal activity of chlorhexidine diacetate (1×10^{-5} mol l^{-1}) towards washed suspensions of *Escherichia coli*, grown in a chemostat under conditions of (a) magnesium limitation (●) and nitrogen limitation (o) and (b) phosphorus limitation (●) and carbon limitation (o) at various specific growth rates. (Reprinted, with permission of Blackwell Scientific Publications, from Wright and Gilbert (1987a))

limitation. The results were not consistent with any simple model for chlorhexidine binding and action and probably reflected a subtle involvement in chlorhexidine permeation through the envelope, and binding to the cell membrane, phospholipid–lipopolysaccharide complexes and cations.

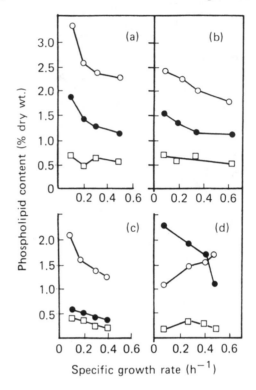

Figure 10.6 Interrelation of phosphatidylglycerol (●), phosphatidylethanolamine (○) and diphosphatidylglycerol (□) content of *Escherichia coli* cells grown in a chemostat under conditions of (a) carbon limitation, (b) nitrogen limitation, (c) phosphate limitation and (d) magnesium limitation at various specific growth rates. (Reprinted, with permission of Blackwell Scientific Publications, from Wright and Gilbert (1987a))

In a related study with the same organism, Wright and Gilbert (1987b) investigated the effects of growth rate and nutrient limitation upon the activity of a homologous series of *n*-alkyltrimethylammonium bromides (C_nTABs) against *E. coli*. Growth inhibitory and bactericidal activities of these compounds are parabolically related to the *n*-alkyl chain length of these compounds and thereby to compound lipophilicity (log P) (Al-Taae *et al.* 1986). The chain length at which optimal activity is demonstrated varies between different cell types and reflects the lipophilicity and barrier properties of the cell envelopes (Hansch and Clayton 1973). Wright and Gilbert (1987b) argued that alteration in envelope-lipophilicity through changes in growth rate and nutrient-limitation (Figs 10.6–10.8), might be expected to produce changes in log P_0 and also in the degree of activity demonstrated by the optimally active compound. Fig. 10.9 shows the effect of *n*-alkyl chain length and growth rate upon the sensitivity of cells prepared under four nutrient-limitations and Fig. 10.10 shows the effects of nutrient-limitation and growth rate upon the activity of one compound, cetrimide (C_{16}TAB). In all cases resistance maximized at growth rates between 0.1 h^{-1} and 0.23 h^{-1} and decreased markedly at the faster growth rates. The compounds chosen represented one side of a parabolic relationship between log P and biological activity where, for cells grown in nutrient broth, activity maximized for the compound with an *n*-alkyl chain length of 16 (Cetrimide

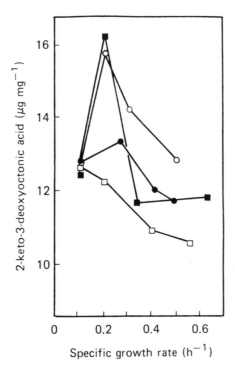

Figure 10.7 Variation in the amounts of lipopolysaccharide as indicated by ketodeoxy-octonic acid level, with growth rate and nutrient limitation for *Escherichia coli* cultures: ■, nitrogen limitation; ○, carbon limitation; □, phosphate limitation; ●, magnesium limitation. (Reprinted, with permission of Blackwell Scientific Publications, from Wright and Gilbert (1987a))

USP). Decreased activity for all the compounds over the slow rates of growth (0.05–0.2 h^{-1}) therefore suggested an overall increase in envelope lipophilicity followed, as growth rate increased, by a steady decrease. The effects upon activity would be expected, and were observed, to be greatest for those compounds with log P closest to log P_0 (C$_{16}$TAB). Whilst similar trends were observed for all four nutrient limitations, carbon-limited cultures were observed to be the most resistant to the agents and showed a tenfold variation in sensitivity, whilst cells grown under phosphorus-limitation were most sensitive and showed an approximate 1000-fold change in sensitivity. The results of this study therefore supported the hypothesis that growth rate and nutrient limitation alter the overall lipophilicity of the cell envelope and thereby influence the optimal value of log P required by compounds in order to traverse it.

10.3 Biofilms, microcolonies and the glycocalyx

Often, in nature, micro-organisms grow in close association with solid surfaces whether these be of soft tissues, *in vivo*, or on nutritionally inert surfaces associated with pipework, submerged particulates or the sides of product containers (Costerton *et al.* 1987).

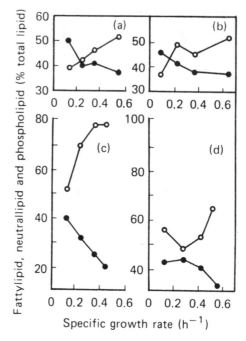

Figure 10.8   Interrelation of phospholipid (●) and fatty and neutral lipid (○) content of *Escherichia coli* cells grown in a chemostat under (a) carbon limitation, (b) nitrogen limitation, (c) phosphorus limitation and (d) magnesium limitation at various specific growth rates. (Reprinted, with permission of Blackwell Scientific Publications, from Wright and Gilbert (1987a))

Association with solid surfaces leads to the generation of consortia of organisms, organized within extensive exopolymer glycocalices, and known as biofilms (Costerton *et al.* 1978). Organization as biofilms is thought to confer many advantages upon the component organisms. These include the provision and maintenance of appropriate physicochemical environments for growth and survival. In this respect, whilst the glycocalyx can function as an ion-exchange column and exclude large, highly charged molecules, most solutes will equilibrate across it and access the resident population. Diffusion limitation by the glycocalyx together with localized high densities of cells creates gradients across the biofilm. Thus, for biofilms established upon impervious surfaces biological demand for oxygen within them creates microaerophilic conditions within the depth of the film and aerobic conditions at the surface. In a similar fashion secondary metabolites, pH and nutrient gradients are also likely to be maintained across the thickness of the biofilm. Cells at different parts of the biofilm are likely to experience different nutrient and physicochemical environments which will in turn affect their physiology. Such physiological variation will in turn affect preservative susceptibility (section 10.2.4). Growth rates (section 10.2.4.2) are likely to be slow in the depths of biofilms and increase with the nutrient gradient.

The ubiquitous nature of the biofilm mode of growth and its importance in microbial ecology is no longer in dispute nor, unfortunately, is their recalcitrance towards conventional biocide treatments and preservative agents. Questions remain, however, as to the nature of such resistance mechanisms and potential methods for their control.

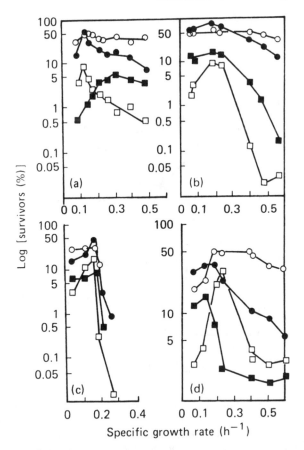

Figure 10.9   Steady-state viability (30 min) following exposure to C_{12}TAB (26 mM) (o), C_{14}TAB (12 mM), (●), C_{16}TAB (4.1 mM) (□) and C_{18}TAB (3.8 mM) (■) of *Escherichia coli*, previously grown in a chemostat under (a) magnesium limitation, (b) nitrogen limitation, (c) phosphate limitation and (d) carbon limitation at various growth rates. (Reprinted, with permission of *Journal of Pharmacy and Pharmacology*, from Wright and Gilbert (1987b))

10.3.1   *Mechanisms of resistance in biofilms*

A number of different approaches have been made in recent years in an attempt to explain and to counter the recalcitrance of microbial biofilms towards chemical and antibiotic treatments (Brown and Gilbert 1993). The major hypotheses are:

(i)   that the glycocalyx excludes and/or influences the access of antimicrobial agents to the underlying organisms,

(ii)   that for chemically reactive antimicrobials such as the halogens and sulphydryl-interactive agents (e.g. isothiazolones) and for physically adsorbed agents active in low concentrations, the surface regions of the glycocalyx and outlying cells react with and quench the biocide,

(iii) that limited availability of key nutrients within the biofilm forces a slowing of the specific growth rate and adoption of phenotypes atypical of planktonic cells exposed to the same growth medium. Heterogeneity through the depth of the biofilm

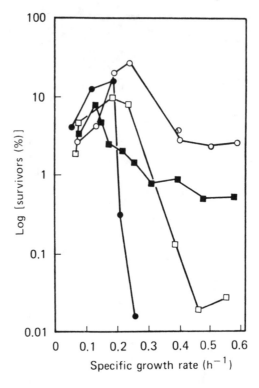

Figure 10.10 Steady-state viability (30 min) following exposure to C_{16}TAB (4.1 mM) of *Escherichia coli*, previously grown in a chemostat under magnesium limitation (■), nitrogen limitation (□), phosphate limitation (●) and carbon limitation (○) at various growth rates. (Reprinted, with permission of *Journal of Pharmacy and Pharmacology*, from Wright and Gilbert (1987b))

    might in this instance lead to a dominance of relatively dominant cells at the base of the biofilm, particularly where the substratum is nutritionally inert,

(iv)   that attachment to surfaces causes the cells to derepress/induce genes associated with a sessile existence which co-incidentally affect antimicrobial susceptibility.

    Central to all of these hypotheses is the principle that attached cells within a biofilm differ significantly from their planktonic, 'domesticated', laboratory counterparts. Most antimicrobial systems have been developed and optimized in laboratories employing planktonic broth cultures as inocula, at least in the initial screening stage. Consequently the possibility must not be overlooked that resistance in biofilms reflects inappropriate selection of compounds for development during the screening process rather than innate recalcitrance of biofilms to antimicrobial agents. Whilst explanations for the observed lack of sensitivity towards today's antimicrobial agents may well be explicable in physi-ological terms, such propositions cannot be disregarded until primary screening for anti-microbial activity routinely involves realistic, relevant challenge systems.

    The contribution of biofilms to preservative resistance within pharmaceutical formula-tions has received little direct study, yet their contribution towards the recalcitrance of naturally occurring micro-organisms is potentially enormous and could influence that of bioburdens introduced into pharmaceutical products from their manufacturing environments. In this respect pseudomonads isolated as biofilms and associated with pipework for the manufacture of povidone-iodine antiseptics (Bond *et al.* 1983) have been implicated in bacteraemias following the use of such contaminated products (Craven *et al.* 1981, Anderson *et al.* 1983).

10.3.1.1 Capsules, slime layers and exopolymers

The outermost extracellular slime layers, or capsules, are generally carbohydrate in nature, their occurrence is often subject to culture conditions and, when present, they facilitate aggregation of cells into microcolonies and biofilms (Gilbert *et al.* 1993). The presence of such materials also confers advantages upon the cells with respect to the colonization of surfaces and resistance to desiccation (Brown and Gilbert 1993, Costerton *et al.* 1987). We may regard these extended layers of glycoprotein, polysaccharides, and other extracellular polymeric substances as surrounding the cells and forming additional barriers to drug entry. In these instances incoming drug molecules, including preservatives, must saturate all unoccupied binding sites within the glycocalyx before reaching the outer and inner membranes. Thus the glycocalyx functions as an ion-exchange column, removing strongly charged molecules from solution as they pass through it (Wagman *et al.* 1975). Growth of cells as biofilms therefore contributes greatly to the resistance of populations to chemical antimicrobial agents. Isolation and subsequent growth of these organisms in typical laboratory culture is then associated with a loss of resistance characteristics. In many situations, such as water storage tanks, cooling towers, air conditioners, sinks, fuel-lines and pipework, it is now well documented that the contaminant microorganisms are present as adhesive biofilms (Marrie and Costerton 1984, Nickel *et al.* 1985) and that these have a significantly increased level of resistance to chemical antimicrobial agents, such as the isothiazolones (Costerton and Lashen 1984), and quaternary ammonium preservatives (Ruseska *et al.* 1982, Stickler *et al.* 1989, Evans *et al.* 1990).

10.3.1.2 Adherence phenotypes

Surface-induced stimulation of bacterial activity has often been observed (Zobell 1943, Fletcher 1985, 1986). There is still much debate, however, as to whether such stimulation results from derepression/induction, through contact with the surface of specific genes, or whether it is indirect and simply reflects physicochemical influences caused by the proximity of a surface (van Loosdrecht *et al.* 1990). Such influences include the accumulation of many substances at surfaces making them available in increased abundance to the organism (Power and Marshall 1988). It has been found that degradation of substrate is enhanced when the degradative organisms are attached to inert surfaces (McFeters *et al.* 1990). This suggests increased production of degradative enzymes by attached bacteria. Similarly, surface-gliding bacteria lack extracellular polymer when grown in planktonic culture (Humphrey *et al.* 1979, Abbanat *et al.* 1988).

Specific gene switch-on has been demonstrated (Belas *et al.* 1986, McCarter *et al.* 1988) in vibrio as they associate with surfaces. Work by Dagostino *et al.* (1991) demonstrated that mutants of an engineered *E. coli* strain expressed a marker gene only when the cells were attached to a polystyrene surface, and not when grown on agar, or in liquid culture. Clearly the possession of specific adherence phenotypes will influence the degradative capacity of spoilage organisms as they attach to product containers and will also influence their susceptibility towards preservatives (Gilbert *et al.* 1987).

10.4 Conclusions

The bacterial cell is remarkably flexible in its structure and composition and will alter in response to changes in the growth environment, particularly the nature of the growth-limiting nutrient and the growth rate. Typical nutrient broth grown cells do not represent the phenotypes expressed in nature. Whilst the changes in preservative sensitivity associated with altered cell-envelope composition are well documented, often too little is known about the mechanism of action of the compounds or their mode of access to the cytosol and cytoplasmic membrane to predict the resistance status from an analysis of envelope composition. It is clear from recent studies that unambiguous relationships between particular envelope components and preservative binding and activity are unlikely to exist and that preservative permeation through the envelope probably involves complex interactions between all the envelope components and the agent.

References

ABBANAT, D. R., GODCHAUX, W., and LEADBETTER, E. R. (1988) Surface induced synthesis of new sulphonolipids in the gliding bacterium *Cytophaga johnsonae*. *Arch. Microbiol.* **120** 358–364.

AL-HITI, M. M. A., and GILBERT, P. (1980) Changes in preservative sensitivity of the USP antimicrobial agents effectiveness test micro-organisms. *J. Appl. Bacteriol.* **49** 119–126.

AL-TAAE, A. N., DICKINSON, N. A., and GILBERT, P. (1986) Antimicrobial activity and physicochemical properties of some alkyltrimethyl ammonium bromides. *Lett. Appl. Microbiol* **1** 101–105.

ANDERSON, R. L. BERKELMAN, R. L., and HOLLAND, B. W. (1983) Microbiologic investigations with iodophor solutions. *Proc. Int. Symp. Povidone*, University of Kentucky Press, Lexington, KY, pp. 146–157.

BASSFORD, P. J., DIEDRICH, D. L., SCHNAITMAN, C. A., and REEVES, P. (1977) Outer-membrane proteins of *Escherichia coli*. VI Protein alterations in bacteriophage resistant mutants. *J. Bacteriol.* **131** 608–622.

BELAS, R., SIMON, M., and SILVERMAN, M. (1986) Regulation of lateral flagella gene transcription in *Vibrio parahaemolyticus J. Bacteriol.* **167** 210–218.

BENZ, R. (1985) Porin from bacterial and mitochondrial outer-membranes. *Crit. Revs. Biochem.* **254** 1457–1461.

BENZ, R., and HANCOCK, R. E. W. (1981) Properties of large ion-permeable pores from protein F of *Pseudomonas aeruginosa* in lipid bilayer membranes. *Biochem. Biophys. Acta* **646** 298–308.

BENZ, R., SCHMID, A., and HANCOCK, R. E. W. (1985) Ionic selectivity of Gram-negative bacterial porins. *J. Bacteriol.* **162** 722–727.

BOGGIS, W., KENWARD, M. A., and BROWN, M. R. W. (1979) Effects of divalent metal cations in the growth medium upon the sensitivity of batch grown *Pseudomonas aeruginosa* to EDTA or Polymyxin A. *J. Appl. Bacteriol.* **47** 477–488.

BOND, W. W., FAVERO, M. S., PETERSON, N. J., and COOK, E. J. (1983) Observations on the intrinsic bacterial contamination of iodophor germicides. *Proc. Ann. Mtg. Amer. Soc. Microbiol.* Q-133.

BROWN, M. R. W. (1975) The role of the envelope in resistance. In: *Resistance of Pseudomonas aeruginosa*, M. R. W. Brown (ed.), John Wiley, Chichester, pp. 71–107.

(1977) Nutrient depletion and antibiotic susceptibility. *J. Antimicrob. Chemother.* **3** 198–201.

BROWN, M. R. W., and GILBERT, P. (1993) Sensitivity of biofilms to antimicrobial agents. *J. Appl. Bacteriol. (Suppl.)* **74** 87s–97s.

BROWN, M. R. W., and MELLING, J. (1969a) Loss of sensitivity to EDTA by *Pseudomonas aeruginosa* grown under conditions of magnesium limitation. *J. Gen. Microbiol.* **54** 263–274.

(1969b) Role of divalent cations in the action of polymyxin B and EDTA on *Pseudomonas aeruginosa. J. Gen. Microbiol.* **59** 263–274.

BROWN, M. R. W., and WILLIAMS, P. (1985) The influence of environment on envelope properties affecting survival of bacteria in infections. *Ann. Revs. Microbiol.* **39** 527–556.

BROWN, M. R. W., ALLISON, D. G., and GILBERT, P. (1988) Resistance of bacterial biofilms to antibiotics: a growth rate related effect. *J. Antimicrob. Chemother.* **22** 777–780.

BROWN, M. R. W., COLLIER, P. J., and GILBERT, P. (1990) Influence of growth rate on the susceptibility to antimicrobial agents: modification of the cell envelope and batch and continuous culture studies. *Antimicrob. Ag. Chemother.* **34** 1623–1628.

BROXTON, P., WOODCOCK, P. M., and GILBERT, P. (1984) Interaction of some polyhexamethylene biguanides and membrane phospholipids in *Escherichia coli. J. Appl. Bacteriol.* **57** 115–124.

BULLMAN, R. A., and STRETTON, R. J. (1975) Influence of 3-chloropropane-1,2-diol on microbial lipid composition and its effect on the activity of some quaternary ammonium disinfectants. *Microbios* **12** 7–27.

COSTERTON, J. W. (1977) Cell-envelope as a barrier to antibiotics. In: *Microbiology.* American Society for Microbiology, Washington, DC, pp. 151–157.

COSTERTON, J. W., and LASHEN, E. S. (1984) Influence of biofilm on efficacy of biocides upon corrosion causing bacteria. *Materials Performance* **23** 34–37.

COSTERTON, J. W., INGRAM, J. M., and CHENG, K.-J. (1974) Structure and function of the cell envelope of Gram-negative bacteria. *Bact. Revs.* **38** 87–110.

(1978) How bacteria stick. *Sci. Amer.* **238** 86–95.

COSTERTON, J. W., CHENG, K.-J., GEESEY, G. G., LADD, P. I., NICKEL, J. C., DASGUPTA, M., and MARRIE, T. J. (1987) Bacterial biofilms in nature and disease. *Ann. Revs. Microbiol.* **41** 435–464.

CRAVEN, D. E., MOODY, B., CONNOLY, M. G., KOLLISCH, N. R. STOTTMEIER, K. D., and McCABE, W. R. (1981) Pseudobacteraemia caused by povidone-iodine antiseptics contaminated by *Pseudomonas aeruginosa. New Eng. J. Med.* **305** 621–623.

DAGOSTINO, L., GOODMAN, A. E., and MARSHALL, K. C. (1991) Physiological responses induced in bacteria adhering to surfaces. *Biofouling,* **4** 113–119.

DEAN, A. C. R. (1972) Influence of environment on the control of enzyme synthesis. *J. Appl. Chem. Biotech.* **22** 245–259.

DEAN, A. C. R., ELLWOOD, J., MELLING, J., and ROBINSON, A. (1976) The action of antibacterial agents on bacteria grown in continuous culture. In: Dean, A. C. R., Ellwood, D. C., Evans, C. G. T., and Melling, J. (eds), *Continuous culture 6: applications and new fields.* Ellis Horwood, Chichester, pp. 251–261.

DECAD, G. M., and NIKAIDO, H. (1976) The outer-membrane of Gram-negative bacteria. Molecular-sieving function of cell-wall. *J. Bacteriol.* **128** 325–336.

DE LA ROSA, E. J., DE PEDRO, M. A., and VASQUEZ, D. (1982) Modification of penicillin binding proteins of *Escherichia coli* associated with changes in the state of growth of cells. *FEMS Microb. Lett.* **14** 91–94.

DE PAMPHILUS, M. C. (1971) Dissociation and reassembly of *Escherichia coli* outer membrane and of lipopolysacharide and their reassembly onto flagellar basal bodies. *J. Bacteriol.* **105** 1184–1199.

ELLWOOD, D. C., and TEMPEST, D. W. (1972) Effects of environment on bacterial wall content and composition. *Adv. Microb. Physiol.* **7** 83–117.

EUDY, W. W., and BURROUGHS, S. E. (1973) Generaton times of *Proteus mirabilis* and *Escherichia coli* in experimental infections. *Chemotherapy* **1** 161–170.

EVANS, D. J., ALLISON, D. G., BROWN, M. R. W., and GILBERT, P. (1990) Effect of growth rate on resistance of Gram-negative biofilms to cetrimide. *J. Antimicrob. Chemotherap.* **26** 473–478.

FINCH, J. E., and BROWN, M. R. W. (1975) The influence of nutrient-limitation in a chemostat on the sensitivity of *Pseudomonas aeruginosa* to polymyxin and EDTA. *J. Antimicrob. Chemother.* **1** 379–386.

FLETCHER, M. (1985) Effect of solid surfaces on the activity of attached bacteria. In: Savage, D. C., and Fletcher, M. (eds), *Bacterial adhesion: mechanisms and physiological significance.* Plenum Press, New York, pp. 339–362.

—— (1986) Measurement of glucose utilisation by *Pseudomonas fluorescens* that are free-living and that are attached to surfaces. *Appl. Environ. Microbiol.* **52** 672–676.

GERHARDT, P., and JUDGE, J. A. (1964) Porosity of isolated cell walls of *Saccharomyces cerevisiae* and *Bacillus megaterium. J. Bacteriol.* **87** 945–995.

GILBERT, P., and BROWN, M. R. W. (1978a) Effect of R-plasmid RP1 and nutrient-depletion upon the gross cellular composition of *Escherichia coli* and its resistance to some uncoupling phenols. *J. Bacteriol.* **133** 1062–1065.

—— (1978b) Influence of growth rate and nutrient limitation on the gross cellular composition of *Pseudomonas aeruginosa* and its resistance to 3- and 4-chlorophenol. *J. Bacteriol.* **133** 1066–1072.

—— (1980) Cell wall mediated changes in the sensitivity of *Bacillus megaterium* towards chlorhexidine and 2-phenoxyethanol associated with growth rate and nutrient limitation. *J. Appl. Bact.* **48** 223–230.

GILBERT, P., and WRIGHT, N. E. (1986) Non-plasmidic resistance towards preservatives in pharmaceutical and cosmetic products. In: Board, R. G., and Allwood, M. C. (eds), *Preservatives in the food, pharmaceutical and cosmetics industries.* Society for Applied Bacteriology Technical Series, Academic Press, London. pp. 155–279.

GILBERT, P., BROWN M. R. W., and COSTERTON, J. W. (1987) Inocular for antimicrobial sensitivity testing: a critical review. *J. Antimicrob. Chemother.* **20** 147–154.

GILBERT, P., COLLIER, P. J., and BROWN, M. R. W. (1990) Influence of growth rate on susceptibility to antimicrobial agents: biofilms, cell cycle, dormancy and stringent response. *Antimicrob. Ag. Chemother.* **34** 1865–1868.

GILBERT, P., EVANS, D. J., and BROWN, M. R. W. (1993) Formation and dispersal of bacterial biofilms, *in vivo* and *in situ. J. Appl. Bacteriol. (Suppl.)* **78** 67s–78s.

GILLELAND, H. E., Jr., STINNETT, J. D., and EAGON, R. G. (1974) Ultrastructural and chemical alteration of the cell envelope of *Pseudomonas aeruginosa* associated with resistance to EDTA resulting from growth in a Mg^{2+}-deficient medium. *J. Bacteriol.* **117** 302–311.

HANCOCK, R. E. W. (1984) Alterations in outer-membrane permeability. *Ann. Revs. Microbiol.* **38** 237–264.

HANCOCK, R. E. W., SCHMID, A., BAUER, K., and BENZ, R. (1986) Role of lysine in ion selectivity of bacterial outer membrane proteins. *Biochim. Biophys. Acta* **860** 263–267.

HANSCH, C. (1973) Quantitative approaches to pharmacological structure–activity relationships. In: Cavallito, C. (ed.), *Structure–activity relationships*, Vol. 1. Pergamon Press, Oxford, pp. 1–21.

HANSCH, C., and CLAYTON, J. M. (1973) Lipophilic character and biological activity of drugs: the parabolic case. *J. Pharm. Sci.* **62** 1–32.

HANSCH, C., and FUJITA, T. (1964) A method for the correlation of biological activity and chemical structure. *J. Amer. Chem. Soc.* **86** 1616–1626.

HANTKE, K. (1976) Phage T6-Colicin K receptor and nucleoside transport in *Escherichia coli. FEBS Lett.* **70** 109–112.

HARDER, K. J., NIKAIDO, H., and MATSUHASHI, M. (1981) Mutants of *Escherichia coli* that are resistant to certain β-lactam compounds lack the OMP-F porin. *Antimicrob. Ag. Chemother.* **20** 549–552.

HERBERT, D. (1956) The continuous culture of bacteria: a theoretical and experimental study. *J. Gen. Microbiol.* **14** 601–622.

HESS, E. L., and LAGG, S. E. (1958) Some physical aspects of the bacterial cell. *Science, NY,* **128** 356–358.

HOLME, T. (1972) Influence of environment on the content and composition of bacterial envelopes. *J. Appl. Chem. Biotech.* **22** 391–399.

HUGO, W. B., and DAVIDSON, J. R. (1973) Effect of cell lipid depletion in *Staphylococcus aureus* upon its resistance to antibacterial drugs. 2. A comparison of the response of normal and lipid-depleted cells of *Staph. aureus* to antibacterial drugs. *Microbios* **8** 73–84.

HUGO, W. B., and FRANKLIN, I. (1968) Cellular lipid and the anti-staphylococcal activity of phenol. *J. Gen. Microbiol.* **52** 365–373.

HUMPHREY, B. A., DICKSON, M. R., and MARSHALL, K. C. (1979) Physiological and *in situ* observations on the adhesion of gliding bacteria to surfaces. *Arch. Microbiol.* **120** 231–238.

IKEDA, T., LEDWITH, A., BAMFORD, C. H., and HANN, R. A. (1984) Interaction of a polymeric biguanide with phospholipid membranes. *Biochim. Biophys. Acta* **769** 57–66.

IMAI, M., INOUE, K., and NOJIMA, S. (1975) Effect of polymyxin B on lysosomal membranes from *Escherichia coli* lipids. *Biochim. Biophys. Acta* **375** 130–137.

KLEMPERER, R. M. M., AL-DUJAILI, D. A., LAWSON, J. J., and BROWN, M. R. W. (1977) The effect of nutrient depletion upon the resistance of *Proteus mirabilis* to antibacterial agents. *J. Appl. Bacteriol.* **43** xvi.

KLEMPERER, R. M. M., ISMAIL, N. T. A. J., and BROWN, M. R. W. (1980) Effect of R-plasmid RP1 and nutrient depletion on the resistance of *Escherichia coli* to cetrimide, chlorhexidine and phenol. *J. Appl. Bacteriol.* **48** 349–357.

LAMBERT, P. A., and HAMMOND, S. M. (1983) Resistance to non-antibiotic antimicrobial agents. In: Hugo, W. B., and Russell, A. D. (eds), *Pharmaceutical microbiology*, 3rd edn. Oxford, Blackwell, pp. 265–274.

LIEN, E. J., HANSCH, C., and ANDERSON, S. M. (1968) Structure–activity correlations for antibacterial agents of Gram-positive and Gram-negative cells. *J. Med. Chem.* **11** 430–441.

LIEVE, L. (1974) The barrier function of the Gram-negative envelope. *Ann. N.Y. Acad. Sci.* **235** 109–129.

LUGTENBERG, B., and VAN ALPHEN, L. (1983) Molecular architecture and functioning of the outer membrane of *Escherichia coli* and other Gram-negative bacteria. *Biochim. Biophys. Acta* **737** 51–115.

MARQUIS, R. E. (1968) Salt-induced contraction of bacterial cell walls. *J. Bacteriol.* **95** 775–781.

MARRIE, T. J., and COSTERTON, J. W. (1984) Scanning and transmission electron microscopy of *in situ* bacterial colonisation of intravenous and intraarterial catheters. *J. Clin. Microbiol.* **19** 687–693.

MCCARTER, L. HILMEN, M., and SILVERMAN, M. (1988) Flagellar dynamometer controls swarmer cell differentiation of *Vibrio parahaemolyticus*. *Cell* **54** 345–351.

MCFETERS, G. A., EGLI, T., WILBERG, E., ADLER, A., SCHNEIDER, R., SNOZZI, M., and GIGER, W. (1990) Activity and adaptation of nitrilotiacetate (NTA) degrading bacteria: field and laboratory studies. *Water Res.* **24** 875–881.

MEADOW, P. M. (1975) Wall and membrane structures in genus *Pseudomonas*. In: Clarke, P. H., and Richmond, M. H. (eds), *Genetics and biochemistry of Pseudomonas*. John Wiley, Chichester, pp. 67–98.

MEERS, J. L., and TEMPEST, D. W. (1970) The influence of growth limiting substrate and medium NaCl concentration on the synthesis of magnesium binding sites in the walls of *Bacillus subtilis* var *niger*. *J. Gen. Microbiol.* **63** 325–331.

MELLING, J., ROBINSON, A., and ELLWOOD, D. C. (1974) Effect of growth environment in a chemostat on the sensitivity of *Pseudomonas aeruginosa* to polymyxin B sulphate. *Proc. Soc. Gen. Microbiol.* **1** 61.

MINNIKIN, D. E., ABDULRAHIMZADEH, H., and BADDILEY, J. (1971) The interrelation of phosphatidylethanolamine and glycosyldiglycerides in bacterial membranes. *Biochem. J.* **124** 447–448.

MITCHELL, P. (1959) Biochemical cytology of microorganisms. *Ann. Revs. Microbiol.* **13** 407–440.

NADIR, M. T., and GILBERT, P. (1980) Influence of potassium chloride upon the binding and antibacterial activity of chlorhexidine diacetate and (ethoxy)5-octyl phenol on *Bacillus megaterium*. *Microbios* **26** 51–63.

NAKAE, T. (1986) Outer-membrane permeability in bacteria. *Crit. Revs. Microbiol.* **19** 145–190.

NAKAE, T., ISHII, J., and TOKUNAGU, M. (1979) Sub-unit structure of functional porin oligomers that form permeability channels in the outer membrane of *Escherichia coli. J. Biol. Chem.* **254** 1457–1461.

NICHOLS, W. W., EVANS, M. J., SLACK, M. P. E., and WALMSLEY, H. L. (1989) The penetration of antibiotics into aggregates of mucoid and non-mucoid *Pseudomonas aeruginosa. J. Gen. Microbiol.* **135** 1291–1303.

NICKEL, J. C., RUSESKA, I., WRIGHT, J. B., and COSTERTON, J. W. (1985) Tobramycin resistance of *Pseudomonas aeruginosa* cells growing as a biofilm on urinary tract catheter materials. *Antimicrob. Ag. Chemother.* **27** 619–624.

NIKAIDO, H. (1976) Outer-membrane of *Salmonella typhimurium*: transmembrane diffusion of some hydrophobic substances. *Biochem. Biophys. Acta* **433** 118–132.

NIKAIDO, H., and NAKAE, T. (1979) The outer-membrane of Gram-negative bacteria. *Adv. Microb. Physiol.* **20** 163–250.

NIKAIDO, H., and VAARA, M. (1985) Molecular basis of bacterial outer-membrane permeability. *Microbiol. Revs* **49** 1–32.

OSBORN, M. J., and WU, H. C. P. (1980) Proteins of the outer-membrane of Gram-negative bacteria. *Ann. Revs Microbiol.* **34** 369–422.

PAYNE, J. W., and GILVARG, C. (1968) Size restrictions on peptide utilisation in *Escherichia coli. J. Biol. Chem.* **243** 6291–6299.

PECHEY, D. T., YAU, A. O. P., and JAMES, A. M. (1974) Total and surface lipids of cells of *Pseudomonas aeruginosa* and their relationship to gentamycin resistance. *Microbios* **11** 77–86.

POWER, K., and MARSHALL, K. C. (1988) Cellular growth and reproduction of marine bacteria on surface-bound substrate. *Biofouling* **1** 163–174.

ROANTREE, R. J., KUO, T.-T., and MACPHEE, D. G. (1977) The effect of defined lipopolysaccharide core defects upon antibiotic resistances of *Salmonella typhimurium. J. Gen. Microbiol.* **103** 223–224.

ROSENBUSCH, J. P. (1974) Characterisation of the major envelope protein from *Escherichia coli*: regular arrangement on the peptidoglycan and unusual dodecylsulphate binding. *J. Biol. Chem.* **249** 8019–8029.

RUSESKA, I., ROBBINS, J., and COSTERTON, J. W. (1982) Biocide testing against corrosion-causing oil-field bacteria helps control plugging. *Oil and Gas Journal* **CS-457** 3–82.

SABET, S. F., and SCHNAITMAN, C. A. (1973) Purification and properties of the colicin E receptor of *Escherichia coli. J. Biol. Chem.* **248** 1797–1806.

SCHERRER, R., and GERHARDT, P. (1971) Molecular sieving by the *Bacillus megaterium* cell wall and protoplast. *J. Bacteriol.* **107** 718–735.

SCHNAITMAN, C. A. (1971a) Solubilisation of the cytoplasmic membrane of *Escherichia coli* by triton X45. *J. Bacteriol.* **108** 545–552.

(1971b) Effect of ethylenediamine tetraacetic acid, Triton X100 and lysozyme on the chemical composition and morphology of isolated cell walls. *J. Bacteriol.* **108** 533–544.

SHAND, G. H., ANWAR, H., KADURUGAMUWA, J., BROWN, M. R. W., SILVERMAN, S. H., and MELLING, J. (1985) *In vivo* evidence that bacteria in urinary tract infection grow under iron restricted conditions. *Infect. Immun.* **48** 35–39.

STICKLER, D., DOLMAN, J., ROLFE, S., and CHAWLA, J. (1989) Activity of antiseptics against *Escherichia coli* growing as biofilms using a physical model of the catheterised bladder. *Brit. J. Urol.* **60** 413–418.

TAMAKI, S., SATO, T., and MATSUHASHI, M. (1971) Role of lipopolysaccharide in antibiotic resistance and bacteriophage adsorption of *Escherichia coli* K12. *J. Bacteriol.* **105** 968–975.

TEMPEST, D. W., DICKS, J. W., and ELLWOOD, D. C. (1968) Influence of growth conditions on the concentration of potassium in *Bacillus subtilis* var *niger* and its positive relationship to cellular ribonucleic acid, teichoic acid and teichuronic acid. *Biochem. J.* **106** 237–243.

TEUBER, M., and BADER, J. (1976) Action of polymyxin B on bacterial membranes: Phosphatidylglycerol and cardiolipin induced susceptibility to polymyxin B in *Achloplasma laidlawii. Antimicrob. Ag. Chemother.* **9** 26–35.

TUOMANEN, E., COZENS, R., TOSCH, W., ZAK, O., and TOMASZ, A. (1986) The rate of killing of *Escherichia coli* by β-lactam antibiotics is strictly proportional to the rate of bacterial growth. *J. Gen. Microbiol.* **132** 1297–1304.

TURNOWSKY, F., BROWN, M. R. W., ANWAR, H., and LAMBERT, P. A. (1983) Effect of iron-limitation of growth rate on the binding of pencillin G to the penicillin binding protein of mucoid and non-mucoid *Pseudomonas aeruginosa. FEMS Lett.* **17** 243–245.

VAN LOOSDRECHT, M. C. M., LYKLEMA, J., NORDE, W., and ZINDER, A. J. B. (1990) Influence of interfaces on microbial activity. *Microbiol. Revs* **54** 75–87.

WAGMAN, G. H., BAILEY, J. V., and WEINSTEIN, M. J. (1975) Binding of aminoglycoside antibiotics to filtration materials. *Antimicrob. Ag. Chemother.* **7** 316–319.

WALLHAEUSSER, K. H. (1984) Antimicrobial preservatives used by the cosmetics industry. In: Kabara, J. J. (ed.), *Cosmetics and drug preservation: principles and practice.* Marcel Dekker, New York, pp. 605–745.

WEINBERG, E. D. (1984) Iron with-holding, a defence against infection and neoplasia. *Physiol. Rev.* **64** 65–102.

WILLIAMS, P., BROWN, M. R. W., and LAMBERT, P. A. (1984) Effect of iron deprivation on the production of siderophores and outer-membrane proteins in *Klebsiella aerogenes. J. Gen. Microbiol.* **130** 2357–2365.

WOOKEY, P., and ROSENBERG, H. (1978) Involvement of inner and outer-membrane components in the transport of iron and colicin B action in *Escherichia coli. J. Bacteriol.* **133** 661–666.

WRIGHT, N. E., and GILBERT, P. (1987a) Influence of specific growth rate and nutrient limitation upon the sensitivity of *Escherichia coli* towards chlorhexidine diacetate. *J. Appl. Bacteriol.* **82** 309–314.

(1987b) Antimicrobial activity of *n*-alkyltrimethyl ammonium bromides: Influence of specific growth rate and nutrient limitation. *J. Pharm. Pharmacol.* **39** 685–690.

(1987c) Influence of specific growth rate upon the sensitivity of *Escherichia coli* towards polymyxin B. *J. Antimicrob Chemother.* **20** 303–312.

YU, F., ICHIHARA, S., and MIZUSHIMA, S. (1979) A major outer-membrane protein (0-8) of *Escherichia coli* K12 exists as a trimer in sodium dodecylsulphate solution. *FEBS Lett.* **100** 71–74.

ZIMMERMAN, W., and ROSSELET, A. (1977) Function of the outer-membrane of *Escherichia coli* as a permeability barrier to beta-lactam antibiotics. *Antimicrob. Ag. Chemother.* **12** 368–372.

ZOBELL, C. E. (1943) The effect of solid interfaces upon bacterial activity. *J. Bacteriol.* **46** 39–56.

11

Safety evaluation of preservatives

J. HAYDEN

BIOS (Consultancy & Contract Research) Limited, Bagshot, Surrey GU19 5ER

Summary

(1) Preservatives are used because they are toxic to micro-organisms but at the same time they are required to be totally safe for humans using the products into which they are incorporated.

(2) The evaluation of the safety of preservatives is essentially similar to that of any other material. It is therefore necessary to assess every possible aspect of toxicity including acute, subacute and chronic exposure as well as reproductive toxicity, mutagenicity and carcinogenicity. It is also important to consider local toxic reactions and the effects of route of administration and the influence of other components of the formulation.

(3) The regulation of the use of preservatives is discussed in the light of the problems of comparing the safety of a new product with that of materials currently in use for which safety test data may not be available.

11.1 Introduction

Antimicrobial preservatives are used because they are toxic to certain micro-organisms whose pathogenic actions are unacceptable or which cause spoilage or deterioration. It is therefore perhaps not surprising that these preservatives may on occasion prove to be toxic to man as well as to micro-organisms. Nevertheless, those substances which are found useful as preservatives are materials which have a sufficiently large margin between the effective antimicrobial concentration in the products to be preserved and the concentration which is toxic to the consumer.

Preservatives do not have any inherent properties which set them apart from other chemicals as far as the evaluation of their toxicity is concerned. The public is increasingly aware of the risks posed by the use of additives in foods and other products to which they are exposed. Preservatives are seen as just another example of those potentially harmful materials whose use must be carefully controlled in order to avoid serious consequences. This simplistic view ignores an essential difference between preservatives and other additives. Failure to use them in adequate quantities is potentially more serious than the risk of their causing adverse effects. The risks of using any formulation ingredient must be compared with the potential benefits to be obtained.

In order to evaluate the toxicity of any substance, it is necessary to know its acute and cumulative toxicity together with observed 'no effect levels', and also to be aware of its mutagenic, teratogenic and carcinogenic potential. In addition, its local actions may be important as well as its liability to cause hypersensitization. Any new preservative is subjected to a battery of mandatory tests to provide these essential data. Most of the preservatives in common use were introduced before such test programmes became legal requirements, and many indeed are included in the generally recognized as safe (GRAS) category (La Du 1977). The safety of these materials has been established by prolonged exposure to humans without adverse comments having been received by the Food and Drug Administration. This is no longer regarded as an acceptable toxicity test procedure but it is nonetheless arguably valid. It is often not possible to find any published toxicity test data on GRAS materials which permit a realistic comparison with new or alternative preservatives.

11.2   Acute toxicity

Preservatives are for the most part used at very low concentrations and the question of whether sufficient will be present to result in a toxic reaction does not arise. The margin of safety may not always be quite as wide as it seems and it is also true that not all preservatives are used at low concentrations. Perhaps one of the oldest and most widely used is sodium chloride, a substance which is still considered by most people to be 'perfectly safe' and (correctly) as essential to life. It may be used at high concentrations as a food preservative and is, of course, normally washed out before the food is eaten. The margin between normal consumption and that which may prove hazardous is nevertheless much narrow than is commonly realized. This has been emphasized by reports of toxicity and fatalities following its use as an emetic, a practice no longer recommended (Robertson 1977, Anon. 1985). When the results of even slight overdosage are described, the picture built up is of a compound too dangerous to use! 'Poisoning was characterized by hypertension, cerebral oedema and deterioration in general health.' This description was given for four patients receiving peritoneal dialysis using solutions in which the sodium chloride content was 146–148 mmol/l^{-1} instead of the 141 mmol/l^{-1} intended (Bisson and Bailey 1979).

11.3   Subacute and chronic toxicity

When substances are given repeatedly, the manifestations of toxic action may differ. It is obvious that, if any substance is repeatedly taken at a rate faster than the individual can detoxify or excrete the material, then there is the possibility of accumulating a dose which attains the acute toxic level (Anon. 1974). Alternatively, repeated doses may cause damage which in itself is not immediately hazardous and perhaps gives rise to no obvious effects. For example gastric erosions are known to occur after ingestion of sulphites (Anon. 1976), possibly resulting in some symptomless blood loss. Without further exposure, such lesions heal and the blood loss is replaced. Continuous exposure may exacerbate the condition such that the blood loss exceeds the replacement rate, resulting in anaemia or even more massive blood loss. This is not a problem at normal use levels of sulphite. Some preservatives contain mercury and prolonged administration can result in accumulation of the heavy metal as, for example, in 'mercurialentis' in which mercury is deposited in the lens of the eye, following use of eye preparations preserved with phenyl mercuric nitrate (Abrams and Majzout 1970).

In order to determine the toxicity of the preservatives in preparations, the total exposure to preservatives, i.e. the sum of that included in all food, drink, cosmetics, toiletries and medicines, must be considered. Allowance must be made in judging the maximum permitted levels for this possible total rather than that included in a single product. This can be particularly important in some individuals. The causal relationship between high sodium consumption and high blood pressure (Anon. 1981) and between sugar and dental caries (Yudkin 1972) are well established. Both salt and sugar are widely used in foods, necessitating dietary control in some directions. Further increase in the total intake from pharmaceutical preparations is then potentially hazardous and current practice is to avoid the use of sugar and high sodium levels wherever possible. Fortunately, the preservatives which are most likely to be used in foods, cosmetics, toiletries and pharmaceuticals are the hydroxybenzoates for which accumulation is not known to occur, and which are used at such levels that acute toxic levels are not expected.

11.4 Reproductive toxicity

As part of the general safety testing of all preservatives, consideration is given to possible effects in higher animals, including man, on the fertility of either sex, gestation or post-natal care or on the development of the embryo. No causal relationship has been shown between any of the commonly used preservatives and reproductive toxicity. Levy (1965) suggested that there might be an interaction between salicylates and benzoic acid. He argued that administration of benzoates used as preservatives can raise blood salicylate levels, and Kimmel *et al.* (1971) showed that concurrent administration of benzoic acid and aspirin in rats caused a significant increase in the teratogenic potential of aspirin. This effect was seen only at doses much larger than ever likely to occur in man. Nevertheless, it serves to illustrate that toxic effects are not always directly mediated. This type of interaction may be particularly difficult to detect, as the bulk of the evidence may well suggest that the preservative is not responsible because in most cases no manifestation of toxicity is seen. There are known actions of some other preservatives which come into the same category. The effects of excess or lack of vitamins is known to be teratogenic (Anon. 1979a). The inactivation of thiamine by sulphites might be expected to interfere with gestation even though no teratogenic effect has been described (Hugo and Causeret 1978). Organic mercury has been shown to be responsible for a congenital form of Minamata disease (Harada 1968, Matsumoto *et al.* 1965, Snyder 1971), but only from high levels of methyl mercury in food rather than from mercury-containing preservatives.

11.5 Carcinogenicity and mutagenicity

Longer human life expectancy means that the population will be exposed to all materials, whether chemical or natural, for a longer time. The higher incidence of cancer linked with longer life expectancy gives cause to question whether the use of such materials has in any way been responsible for the higher incidence of cancer or whether the increase is simply the result of longer life-span, allowing time for the cancers to develop. Several preservatives have been subject to close scrutiny as possible carcinogens, and their use regulated. Chlorofom is not permitted in the USA following the demonstration of carcinogenic action in animals and is restricted to concentrations of less than 0.5% in the UK (Anon. 1979b). In this case these authorities have differed in their assessment of the risk. Formaldehyde has also been shown to cause cancers in rats after inhalation (Jensen 1980), and has also been suspected of combining with chlorine to form the carcinogenic bis-chloromethyl ether which is known to be a very potent carcinogen (Drew *et al.* 1975). The long-term safe use of formaldehyde clearly illustrates how effective legislative control of its concentration in use has reduced its toxic potential to a safe level (Anon. 1986a). Another possible cause for concern is the possibility of preservatives reacting with amines or amides to form carcinogenic nitrosamines. This is possible with nitrates and nitrites and therefore theoretically with Bronopol, which is known to form free nitrite when it degrades. In practice, mutagenicity tests were negative and carcinogenicity tests in rats and mice on Bronopol revealed no signs of any effect on tumour incidence (Anon. 1986b, Holland 1981).

Published data on the mutagenicity of preservatives are very limited and testing does pose some problems in recognized bacterial tests, such as the 'Ames' test (Ames *et al.* 1975). These tests can clearly only be carried out at concentrations which are lower than those to which human cells will be exposed.

11.6   Local toxicity

It is apparent, when searching for reports of adverse reactions to preservatives, that the most frequent occurrences are local reactions such as skin irritation or hyper-sensitivity reactions. The current E list of preservative food additives lists 34 different products (Anon. 1975). When adjusted to take account of different salts and minor variations there are really only ten classes (Table 11.1), but half of these are known to cause either local irritation or some form of allergic reaction in susceptible individuals (Hanssen and Marsden 1984). When those preservatives which are reserved for pharmaceuticals, cosmetics and toiletries are added to this list, the situation does not appear to be much better. No fewer than 28 different materials are known to cause adverse skin reactions and a further nine are restricted to use in products which are washed off immediately after use. Of the remaining 23, most are infrequently used and reliable data are not available, leaving only two which are theoretically acceptable for local use (Marzulli and Maibach 1973); in practice, however, they are all used without serious consequences because effective safety testing has determined the maximum concentration and application acceptable for each preservative.

There are two classes of product which make particularly severe demands on the preservative system used. Firstly, products which are used in the eye, most notably those concerned with contact lens care, pose the problem of balancing the risk of local reaction against the need to ensure that they are adequately preserved. Thiomersal appears to be the most suitable preservative although it is by no means ideal (Healey 1982). Products for instillation into the ear must be well preserved but must also be safe if the ear-drum is perforated. At higher than the permitted preservative levels, chlorhexidine, thiomersal and benzalkonium chloride have been incriminated with causing permanent damage to the inner ear (Honigman 1975). This effect can be demonstrated in the laboratory using an electrophysical technique (Harpur 1981).

Table 11.1 Permitted preservatives. The preservatives in food regulations (Anon. 1975)

Preservative class	E numbers	Local adverse reactions
Benzoic acid and salts	E210–E213	Allergic reactions
Hydroxybenzoates	E214–E219	Allergic reactions
Biphenyls	E230–E232	Eye and nose irritation
Nitrites and nitrates	E249–E252	Allergic reactions
Propionates	E280–E282	None known
Sorbates	E200–E203	Eye and skin irritation
Sulphites	E220–E227	Respiratory and gastric irritation
Thiabendazole	E233	Skin and gastric irritation
Hexamine	E239	Skin rashes
Nisin	—	None known

11.7 *In vitro* techniques

The use of animals for toxicity testing is a highly emotive issue and much effort continues to be expended in the search for alternative techniques to reduce the numbers of animals used. Refinement of conventional testing techniques has undoubtedly increased the efficiency of testing and resulted in some decrease in animal numbers used. *In vitro* techniques are usually based on the use of isolated organs, tissues or cells in culture or more specific studies on receptor sites or enzyme preparations. These do not of course entirely remove the need for the use of animals but do avoid the more emotive aspects of pain and suffering as a result of prolonged treatments. In the absence of irrefutable validation of such new methods against the more conventional procedures and more importantly, practical experience, there remain serious problems of interpretation. Despite their expressed wish to restrict the use of animals, Regulatory Authorities will require more definite proof of the adequacy of such tests before they will incorporate them into national or international legislation. They are indeed taking active steps such as collaborative studies of the relationship between *in vivo* primary irritation and *in vitro* experimental models (Berlin and van de Venne 1989).

11.8 Effect of route of administration and formulation

It is important that the toxicity of a preservative should be assessed with respect to the actual site of exposure. A substance may be quite unacceptably toxic by one route and yet quite harmless by others. This is perhaps most dramatically illustrated by the cationic surfactants. The list of adverse effects given for cetrimide (Anon. 1993a) appears at first sight to preclude its use anywhere in contact with the body. In fact, on the normal intact skin there is usually no problem, particularly at the concentrations used for preservation. When taken by mouth, it may cause nausea and vomiting and intravenous or intrauterine administration may result in haemolysis. Above all, cetrimide has depolarizing activity typical of quaternary ammonium compounds and if given by injection can produce a whole range of toxic symptoms: convulsions, dyspnoea, cyanosis, hypotension and coma. It can therefore only be used in topically applied products.

Although the route of administration may drastically affect the apparent toxicity of the preservative, the other components of a formulation are also important. Preservatives may become adsorbed onto some constituent of a formulation and as a result cause less effect than might be expected from the absolute quantity present. In contrast, the presence of surface-active agents may alter the permeability of the skin and thus increase the bio-availability of the preservative. This is particularly true if local inflammation occurs. Indeed, some use is made of this effect in the 'maximization' test procedure of Kligman (1966) which is used to assess the propensity of substances to cause delayed hypersensitivity reactions in man.

It is also necessary to consider the consumer. Not only are there differences between individuals for which no immediate explanation can be given but also there may be differences in response related to age (Bowman and Rand 1980), race (Beutler 1971), health (Freedman 1980), and diet (Anon. 1974).

11.9   Regulatory aspects

The use of preservatives is regulated in many countries. In the UK, preservatives are regulated under a variety of headings: pharmaceuticals are controlled by the Department of Health under the Medicines Act 1968, cosmetics by the Department of Trade and Industry under the Cosmetic Product (Safety) Regulations 1984 as well as under relevant EC Directives, and foods by the Ministry of Agriculture, Fisheries and Food under the Preservatives in Food Regulations 1975. In the USA the situation is simpler in that a single authority, the Food and Drug Administration, is responsible for all three classes. In the EC, control of the use of preservatives in cosmetics and toiletries was defined in Directive 76/768/EEC which has been modified several times since, most recently by 92/86/EEC (Anon. 1993b). Pharmaceuticals may be registered using the multistate procedure of Directive 83/570/EEC or under individual national legislation. Although there may be considerable differences in specific cases, the overall effect has been that many preservatives are no longer permitted and others have become more and more restricted in use. This regulation is usually because of their potential for toxicity rather than their inadequacy as preservatives. Reference has been made above to the restrictions placed on the use of chloroform; the Food and Drug Administration banned it completely in 1976, and the UK restricted its use in 1979 to a maximum of 0.5% w/w or v/v in products to be taken internally but otherwise permitting up to 4% in toothpaste and no limit in topical products. The EC has progressively restricted its use until it attained its current status of being totally banned from cosmetic products, being transferred to Annex II, 'List of substances which cosmetic products must not contain', by Directive 86/179/EEC of 28 February 1986 (Anon. 1986a). These differences reflect the alternative views which authorities may taken when faced with the same problem.

This regulatory trend is by no means new. The *British Medical Journal* in 1887 (Anon. 1987) reported on the adulteration of milk with preservatives carried out by dealers to avoid wastage of milk which they were unable to sell while still fresh. Various proprietary nostrums containing boric and salicylic acids were added at levels which could result in the presence of obvious crystals in butter made from such milk. As a result, Germany, France and parts of the USA took action to outlaw such practices.

A final word is in order concerning the regulatory situation when a novel preservative substance is used. Much depends on the type of product and the country concerned but perhaps the most critical appraisal is reserved for a preservative to be used in a medicine. As a general rule, such an additive would be subjected to the same evaluation of potential risk against potential benefit as would be applied to a new drug substance. On the risk side, the substance would be evaluated according to the guidelines laid down for a new drug and evidence would be expected that would convince the authorities that it did not present an unacceptable hazard. On the efficacy side, it obviously has to be able to achieve satisfactory preservative efficacy. In this latter respect, it will clearly be judged against existing materials whose advantages and disadvantages are well established. Even though the weight of evidence may appear to favour the new introduction, caution is always exercised; many of the adverse reactions attributed to existing preservatives are extremely rare events and in some cases appear to be based on quite inadequate proof. It is equally undesirable that new materials should be ignored, because it has not been possible to show that they would be superior to existing materials.

Unfortunately, the cost of providing evidence of safety for a new material is such as to provide little apparent commercial incentive to justify the risk involved in such an invest-ment.

11.10   Conclusions

In most products, be they foods, cosmetics, toiletries or medicines, preservatives are present at the lowest effective levels and their possible toxic hazards are not therefore the most immediate cause for concern. Although these low levels usually provide the greatest reassurance that preservative toxicity is not a problem, it cannot be assumed that an assessment of the hazard is unnecessary. Furthermore, it also means that toxic effects resulting from the preservatives and which are not detected by safety testing during development will be less readily identified and their true cause may not be immediately realized.

None of the preservatives in current use can be regarded as ideal. They vary in both efficacy and safety, and the relevance of their particular properties varies according to the attributes of the product. It follows that it is not possible to identify an ideal preservative even for a particular type of product. In each case, in deciding which preservation system to use, it is necessary to take into consideration the benefit to be gained and to balance this against not only the risks of inadequate preservation but also the possibility of adverse reactions resulting from the preservatives used. In all cases the safety of a prod-uct must be assessed and suitable tests performed to ensure preservative safety in use.

References

ABRAMS, J. D., and MAJZOUT, V. (1970) Mercury content of the human lens. *Br. J. Ophthalmia* **54** 59–61.

AMES, B. N., McCANN, J., and YAMASAKI, E. (1975) Methods for detecting carcinogens and mutagens with the Salmonella/mammalian-microsome mutagenicity test. *Mutat. Res.* **31** 347–364.

ANON. (1974) *WHO Chron.* **28** 8.

(1975) Statutory Instrument 1975/1487, *The preservatives in food regulations.* HMSO, London.

(1976) Evaluation of the health aspects of sulfiting agents as food ingredients, FDA Report No. BB-265-508. Food and Drugs Administration.

(1979a) 17th Report of Joint FAO–WHO Expert Committee on Food Additives, WHO Technical Report Series No. 539. Food and Agriculture Organization–World Health Organization, New York.

(1979b) Statutory Instrument 179/382, *The medicines (chlorofom prohibition) order*. HMSO, London.

(1981) New evidence linking salt and hypertension. *Br. Med. J.* **282** 1993–1994.

(1985) Removal of poison from the stomach. *British National Formulary Number 10*. British Medical Association and Pharmaceutical Society of Great Britain, London, p. 31.

(1986a) Seventh Commission Directive (86/179/EEC), Annex II. *Off. J. Eur. Communities* No. L138, 40.

(1986b) Mutagenicity and carcinogenicity, Bronopol Technical Bulletin No. 5, 24. Boots Co. Ltd, Nottingham.

(1987) One hundred years ago. *Br. Med. J.* **294** 404.

(1993a) In: Martindale (ed.), *The extra pharmacopoeia*. 30th edn. Pharmaceutical Press, London, p. 787.2.

(1993b) Cosmetics permitted ingredients. *Regulatory Affairs Journal* **3** 121.

BERLIN and VAN DER VENNE (ed.) (1989) Collaborative study on relationship between 'in vitro' primary irritation and 'in vivo' experimental models. Health and Safety Directorate. Commission of the European Communities, Luxembourg.

BEUTLER, E. (1971) Abnormalities of the hexose monophosphate shunt. *Semin. Haematol* **8** 311–347.

BISSON, P. G., and BAILEY, K. M. (1979) Sodium in peritoneal dialysis solutions. *Br. Med. J.* **1** 1322–1323.

BOWMAN, W. C., and RAND, M. J. (1980) Agents producing metahaemoglobinaemia. In: *Textbook of pharmacology*. 2nd edn. Blackwell, Oxford, pp. 21–47.

DREW, R. T., LARKIN, S., KUSCHNER, M., and NELSON, M. (1975) Inhalation carcinogenicity of alpha halo ethers. 1. The acute inhalation toxicity of chloromethyl ether and bis (chloromethyl) ether. *Arch. Environ. Health* **30** 61–69.

FREEDMAN, B. J. (1980) Sulphur dioxide in foods and beverages: its use as a preservative and its effect on asthma. *Br. J. Dis. Chest.* **74** 128.

HANSSEN, M., and MARSDEN, J. (1984) *E for additives. The complete E number guide*. Thorsons, Wellingborough.

HARADA, Y. (1968) In: *Minamata Disease*. Kumanoto University Study Group of Minamata Disease, Kumanoto, p. 73.

HARPUR, E. S. (1981) Ototoxicological testing. In: *Testing for toxicity*. Taylor & Francis, London, p. 227.

HEALEY, J. N. C. (1982) A guide to contact lens care. *Pharm. J.* **229** 650–655.

HOLLAND, V. R. (1981) BNPD and nitrosamine formation. *Cosmet. Technol.* **3** 31–36.

HONIGMAN, J. L. (1975) Disinfectant ototoxicity. *Pharm. J.* **2** 523.

HUGOT, D., and CAUSERET, J. (1978) Effects of the ingestion of tannic acid, potassium metabisulphite and ethanol administered alone or in combination on reproduction in the rat. *C.R. Soc. Biol.* **172** 4705.

JENSEN, O. M. (1980) Cancer risk for formaldehyde. *Lancet* **ii** 481–482.

KIMMEL, C. A., WILSON, J. G., and SCHUMACHER, H. J. (1971) Studies on metabolism and identification of the causative agent in aspirin teratogenesis in rats. *Teratology* **4** 15–24.

KLIGMAN, A. (1966) The identification of contact allergens by human assay. III. The maximisation test: a procedure for screening and rating contact sensitizers. *J. Invest. Dermatol.* **47** 393–409.

LA DU, B. N. (1977) Effects of GRAS substances on pharmacologic effects of drugs. *Clin. Pharmacol. Ther.* **22** 743–748.

LEVY, G. (1965) Pharmacokinetics of salicylate elimination in man. *J. Pharm. Sci.* **54** 959–967.

MARZULLI, F. N., and MAIBACH, H. I. (1973) Anti-microbials: experimental contact sensitization in man. *J. Soc. Cosmet. Chem.* **24** 563–574.

MATSUMOTO, H. G., GOYO, K., and TAKAVECHI, T. (1965) Fetal Minamata disease. A neuropathological study of two cases of intrauterine intoxication by a methyl mercury compound. *J. Neuropath. Exp. Neurol.* **24** 563–574.

ROBERTSON, W. O. (1977) Danger of salt as an emetic. *Br. Med. J.* **2** 1022.

SNYDER, R. D. (1971) Congenital mercury poisoning. *New Engl. J. Med.* **284** 1014–1016.

VAUGHAN, J. S., and PORTER, D. A. (1993) A new *in vitro* method for assessing the potential toxicity of soft contact lens care solutions. *CLAO Journal* **19** 54.

YUDKIN, J. (1972) Sugar and disease. *Nature (London)* **239** 197–199.

PART FOUR

Microbiological control: methods and standards

12

Preservative efficacy testing of pharmaceuticals, cosmetics and toiletries and its limitations

A. L. DAVISON

Formerly of St. Bartholomew's Hospital, London EC1A 7BE

Summary

(1)   Preservative efficacy can not be predicted and therefore is determined empirically against a standardized microbial challenge. Activity is dependent on the composition of the ingredients in the formulation and the container.

(2)   Definition of adequacy of preservation is related to the type of formulation and its mode of use. No universally agreed criteria exist owing to the conflicting severity of existing and proposed pharmacopoeial recommendations.

(3)   These are reviewed and factors concomitant with the evolvement of realistic recommendations for different categories of pharmaceutical, cosmetic and toiletry products discussed.

12.1   Introduction

Preservation of pharmaceuticals, cosmetics and toiletries is necessary to prevent formulations from becoming infected with highly resistant and hazardous micro-organisms during their repeated use. There are no internationally agreed standards for microbial preservation owing to the diversity of the microbial insult, which different medicaments may have to combat throughout their repeated use and intervening storage. The preservative efficacy of a formulation cannot be predicted and has to be established empirically by microbial challenge, as the biocidal or biostatic activity of the preservative is dependent on the effect of individual ingredients and the container on partitioning and absorption of preservative.

However, the adequacy of preservation needs to be equated with the potential risk to the user, as infection or colonization with dermal or oral micro-organisms is dependent on the route of application and the susceptibility of the user to infection. Precise definition of effective preservation therefore is difficult, as opinions differ based on individual experience with distinct preservatives and formulations. Consequently, recommendations and guidelines contained in different compendia are conflicting. Those of particular relevance to pharmaceuticals, cosmetics and toiletries, together with their limitations, will be discussed.

12.2 Pharmaceuticals

12.2.1 *Applicability of compendial tests*

Preservative efficacy tests are described in the *British Pharmacopoeia (BP)* (Anon. 1980, 1988, 1993a) the *United States Pharmacopoeia (USP)* (Anon. 1990a, 1995a) and more recently in the *European Pharmacopoeia (Ph. Eur. or EP)* (Anon. 1993b). The *USP* has specific requirements for sterile aqueous injectable, ophthalmic, otic and nasal products which are not applicable to aqueous topical and oral products. These requirements for a slow rate of kill without effective lethality have remained unchanged for 20 years. They are regarded as insufficient in the UK, where the recommendation for a rapid bacterial lethality for injectable preparations was proposed by a Ministry of Health working party in 1957 (Sykes 1958). The *BP* in 1980 introduced specific recommendations for differing categories of preparation, which of necessity were more stringent for sterile than for non-sterile formulations. These were intended as non-mandatory guidelines for formulation development and licensing to provide and evolve a reasonable basis for assessment of expected preservative efficacy consistent with product safety. These guidelines are regarded as too stringent to permit use of the wider range of preservatives employed in some formulations in Europe. Consequently, the *Ph. Eur.* (1993) has evolved less stringent, compromise criteria which replace the *BP* (1993) guidelines and facilitate harmonization of preservative efficacy testing.

12.2.2 *The British Pharmacopoeia and European Pharmacopoeia tests*

The test protocols of the two compendia are now identical (Anon. 1993a, Anon. 1993b); separate samples of the formulation are challenged with designated species of Gram-positive and Gram-negative bacteria, yeast and mould and the viable count of each species then determined at intervals over at least 28 days. *Staphylococcus aureus, Pseudomonas aeruginosa, Escherichia coli, Candida albicans* and *Aspergillus niger* are used to provide a standardized challenge with resilient and pathogenic species capable of ingression both during manufacture or through repeated use and with the potential to adapt and grow in the formulation. This standardized challenge is representative of species which cause dermal and enteric infection, product degradation and spoilage. *E. coli* is used in the context of an enteric pathogen and is only applicable therefore to oral liquids. The designated species should be supplemented, however, with other strains which may present a particular challenge to the preparation and *Zygosaccharomyces rouxii* is included for syrups to indicate that fermentative degradation by osmotolerant yeasts will not occur.

The conditions for preparation of inocula and determination of viable count by enumeration of bacteria on tryptone soya agar and of yeast and mould on Sabouraud dextrose agar are standardized as shown in Table 12.1. The major differences compared with the *USP* are over preparation of inoculum and determination of viable count. Restriction to use of freshly prepared inocula in the *BP/Ph. Eur.* eliminates potential for variation in response concomitant with the alternative use of stored inocula, as permitted in the *USP*. The ideal of a single method of colony-forming enumeration by pour plate, on TSA as in the *USP*, is too restrictive for the vast range of formulations with differing preservative systems. The alternatives of enumeration by spread plate and membrane filtration therefore are included in the *BP* 1993 and *Ph. Eur.* 1993 with use of Sabouraud

Table 12.1 Conditions for preparation of inocula and determination of viable count in the preservative efficacy test of the *British Pharmacopoeia* (Anon. 1993a) and the *European Pharmacopoeia* (Anon. 1993b)

	British Pharmacopoeia (Anon. 1993a) *European Pharmacopoeia* (Anon. 1993b)
Growth of inoculum	
Bacteria	Tryptone soya agar, 30–35°C, 18–24 hr
Yeast	Sabouraud dextrose agar, 20–25°C, 2 days
Mould	Sabouraud dextrose agar, 20–25°C, 7 days
Diluent	0.1% peptone water
Level of inoculum	10^5–10^6 cfu g^{-1} or ml^{-1}
Age of inoculum	Fresh
Plate count	
Medium for bacteria	Tryptone soya agar
Medium for yeast and mould	Tryptone soya agar or Sabouraud dextrose agar
Incubation for bacteria	30–35°C, 3 days
Incubation for yeast and mould	20–25°C, 5 days

dextrose agar for yeast and mould. These techniques have different potential limits of detection: less than 1 for membrane filtration, less than 10 for pour plate and less than 100 for spread plate. Hence the limit of detection needs to be defined for harmonization as, for instance, less than 100 cfu for all techniques. This is not unrealistic, where a challenge of 10^6 is selected to delineate between a bactericidal and bacteristatic effect with effective lethality defined as a 99.99% reduction, or less than 100. Verification of inactivation of preservative by demonstration of a low-level recovery (less than 100) is straightforward with soluble formulations, but problems occur with suspensions where inhibition cannot be removed by membrane filtration and no specific inactivator is available. A realistic solution would be to define no recovery as absence of revivable organisms for situations where verification is not feasible.

12.2.3 *Criteria for parenteral and ophthalmic preparations*

The fundamental distinction between the *BP* 1980 and the *Ph. Eur.* 1993 recommendations, now included in the *BP* 1993 (see Table 12.2), is in the differing severity of the criteria used for assessment of adequacy of preservation. The *BP* 1980 guidelines were based on a rapid bactericidal activity to ensure rapid elimination of pathogens in the interval between repeat use of the preparation (e.g. 4–6 hours for ophthalmics) and where attainability was restricted to rapidly acting preservatives used in the UK. The replacement two-tier system evolved by the *Ph. Eur.* is designed to facilitate use of the wider range of preservatives employed in Europe. The target criteria express the recommended efficacy with the minimum criteria reserved for use in justified cases, where for instance toxicological considerations preclude demonstration of achievement of target criteria.

Investigation by a *BP* working party compared in-use contamination with preservative efficacy in eye-drops containing different preservative systems (Davison *et al.* 1991).

Table 12.2 Criteria of acceptance for preservation of parenteral and ophthalmic preparations

	Log reduction				
	6 hours	24 hours	7 days	14 days	28 days
BP 1980, 1988					
Bacteria	3	NR			NR
Fungi			2		NI
BP 1993					
Ph. Eur. 1993					
Bacteria					
Target A	2	3			NR
Minimum B		1	3		NI
Fungi					
Target A			2		NI
Minimum B				1	NI

NR, no recovery.
NI, no increase.

This confirmed that the target criteria provide a reliable alternative basis to the more stringent *BP* 1980 guidelines for assessment of preservative efficacy. The poor reliability of the minimum criteria was highlighted by the high level (10^3–10^5 ml^{-1}), but low incidence (1%) of in-use contamination with the ocular pathogen *Staphylococcus hominis* which was restricted to formulations containing the slow-acting preservative thiomersal. These passed the minimum criteria, but failed the target criteria (see Table 12.3). This questions the validity of the USP proposals (Anon. 1995b) which are less stringent than the minimum requirements (viz. log 3 bacterial reduction in 14 days against 7 days) as a reliable basis for prediction of adequacy of preservation.

12.2.4 *Criteria for topical preparations*

The fundamental distinction between the *BP* 1988 recommendations, retained in the *BP* 1993, and those published in the *Ph. Eur.* (Anon. 1994) and most recently in the 1993 *BP Addendum* (Anon. 1995b), is in the differing severity of the criteria used for assessment of adequacy of preservation (Table 12.4).

The *BP* recommendations were based on a bactericidal and fungicidal activity necessary to ensure rapid elimination of dermal pathogens, encountered during use. Acceptance of these recommendations for realistic evaluation of preservative efficacy is dependent on their overall attainability and reproducibility with those preservatives most frequently used in creams and lotions, e.g. chlorocresol, phenoxyethanol and parabens. The attainability and reproducibility of the bactericidal recommendation can readily be verified with a cetomacrogol cream, preserved with 0.1% chlorocresol, comparable data being obtained from examination of duplicate samples on different occasions (Table 12.5). The 10^3 bacterial reduction, although borderline at 1 day, is achieved in 2 days with no subsequent proliferation evident at 7 and 28 days. The latter is vital as the classical pattern for inadequate preservation against *P. aeruginosa* is an initial reduction

Table 12.3 Correlation of in-use contamination with preservative efficacy of eye-drops

Preparation (preservative)	Incidence of in-use contamination		Preservative efficacy (Ph. Eur.)	
	$<10^2/ml^a$	$10^3-10^5/ml^b$	Target A	Minimum B
Dexamerhasome (thiomersal)	5/221	4/221	Fail[c]	Pass
Saline (thiomersal)	4/154	3/154	Fail[c]	Pass
Predinisolone (BKC)	9/97	0/97	Pass	Pass
Benoxinate (CA)	6/132	0/132	Pass	Pass

[a] *S. epidermidis S. hominis M. luteus* and aerobic spore bearers.
[b] *S. hominis*
[c] Fail against *S. aureus* but not *P. aeruginosa*.
BKC, benzalkonium chloride.
CA, chlorhexidine acetate.

Table 12.4 Criteria of acceptance for preservation of topical preparations

	Log reduction			
	2 days	7 days	14 days	28 days
BP 1988, 1993				
Bacteria	3	NR		NR
Fungi			2	NI
Ph. Eur. 1994 BP Adden. 1995				
Bacteria				
Target A	2	3		NI
Minimum B			3	NI
Fungi				
Target A			2	NI
Minimum B			1	NI

NR, no recovery.
NI, no increase.

between 1 and 2 days, possibly below the limit of detection followed by adaptive growth and increase to 10^6 in 7–28 days.

A fungicidal recommendation of a 10^2 reduction in 14 days with no increase at 28 days is a realistic criterion, as fungi are not consistently eliminated at 14 and 28 days (Table 12.6).

The two-tier system, similar to that for parenterals and ophthalmics, has now replaced existing *BP* guidelines (Table 12.4) (Anon. 1994, Anon. 1995c). This system with less stringent criteria than the *BP* 1988 is designed to facilitate use of a less restricted range of preservatives with target criteria expressing the recommended efficacy and minimum criteria reserved for use in justified cases.

Table 12.5   Results of the *British Pharmacopoeia* (1988) preservative efficacy test on a cetomacrogol cream, containing 0.1% w/w chlorocresol, against *Staphylococcus aureus* and *Pseudomonas aeruginosa*

Species	Sample	Inoculum	Log number of survivors (by plate count) after the following times			
			1 day	2 days	7 days	28 days
Staphylococcus aureus	IA		1.0	—	—	—
	IB	5.9	3.2	—	—	—
	IIA		2.3	2.0	—	—
	IIB	6.2	3.3	1.6	—	—
Pseudomonas aeruginosa	IA		—	—	—	—
	IB	6.0	—	—	—	—
	IIA		—	—	—	—
	IIB	6.4	—	—	—	—

A and B are duplicate samples; I and II are duplicate tests on separate occasions.
—, less than 10 cfu g^{-1}.

Table 12.6   Results of the *British Pharmacopoeia* (1988) preservative efficacy test on cetomacrogol cream, containing 0.1% w/w chlorocresol, against *Candida albicans* and *Aspergillus niger*

Species	Sample	Inoculum	Log number of survivors (by plate count) after the following times		
			7 days	14 days	28 days
Candida albicans	IA		—	—	—
	IB	6.0	1.3	—	—
	IIA		2.9	3.9	1.5
	IIB	6.1	1.8	—	—
Aspergillus niger	IA		—	—	—
	IB	6.0	—	—	—
	IIA		3.1	2.0	2.5
	IIB	6.1	2.6	2.0	1.6

A and B are duplicate samples; I and II are duplicate tests on separate samples.
—, less than 10 cfu g^{-1}.

A further comparison of in-use contamination with preservative efficacy by a *BP* working party (Spooner and Davison 1993) confirmed that the target criteria provide a reliable, although less stringent, alternative to the *BP* 1988 guideline for assessment of preservative efficacy of topical preparations. Correlation of in use contamination with deficiencies in preservative efficacy is less precise with topicals due to the enhanced protection afforded by certain types of packaging, namely tubes compared with jars; formulations may exhibit contamination in jars not evident in tubes. Higher levels of in use contamination, including hazardous levels of *P. aeruginosa* (10^4–10^5 g^{-1}) and other pseudomonads, were evident in formulations which passed the minimum criteria, but failed the target criteria (see Table 12.7).

A major deficiency of the minimum criteria with examinations only at 14 and 28 days is the inability to detect regrowth of *P. aeruginosa* between 7 and 14 days, indicative of potentially inadequate preservation. This questions the validity of the latest USP proposals (Anon. 1995b) for topicals, which are less demanding than the minimum criteria of the *Ph. Eur.* (Anon. 1994).

Table 12.7 Correlation of in-use contamination with preservative efficacy of topical preparations

Preparation (preservative)	In-use contamination			Preservative efficacy (Ph. Eur.)	
	Incidence	Composition	Mean count/g	Target A	Minimum B
A (BKC cetrimide)	34/65	GNR	5×10^4	Fail[a]	Pass
B (paraben)	8/191	GPC GPR ASB	5×10^2	Fail[a]	Pass
C (chlorocresol)	4/12	GNR GPC ASB	8×10^2	Fail[b]	Pass
D (paraben)	8/34	GPR GPC	1×10^2	Pass	Pass
E (Bronopol, myacide)	2/54	ASB	$<10^2$	Pass	Pass
F (phenoxyethanol)	7/20	GPC ASB	$<10^2$	Pass	Pass

GNR, Gram-negative bacilli; *P. aeruginosa* and other pseudomonads.
GPC, Gram-positive cocci; coagulase-negative staphylococci.
GPR, Gram-positive bacilli; diphtheroids.
ASB, aerobic sporing bacilli.
[a] Fail against *P. aeruginosa*; regrowth between 7 and 14 days.
[b] Fail against *S. aureus*; < log 3 reduction in 7 days.

12.2.5 Criteria for oral liquids

Differences between the *BP* and the *Ph. Eur.* recommendations for oral liquid preparations are shown in Table 12.8. Difficulty with regard to attainability of the *BP* 1980

Table 12.8   Criteria of acceptance for preservation of aqueous oral preparations

	British Pharmacopoeia (Anon. 1980)	*British Pharmacopoeia* 1988	*European Pharmacopoeia* 1993 *British Pharmacopoeia* 1993
Bacteria	10^3 reduction in 2 days No recovery at 7 days, 14 days and 28 days	10^2 reduction in 7 days No increase at 14 days and 28 days	10^3 reduction in 14 days No increase at 28 days
Fungi	10^2 reduction in 14 days No recovery at 28 days	No increase at 14 days and 28 days	10^1 reduction in 14 days No increase at 28 days

criteria resulted in a relaxation in the *BP* 1988; data by Kurup and Wan (1986) illustrated the failure of preparations to meet the former recommendation of a 10^3 bacterial reduction in 2 days. The single-tier systems of the *BP* 1988 and *Ph. Eur.* 1993 are essentially similar, based on a biostatic and fungistat requirement to prevent the possibility of bacterial and fungal spoilage occurring during sue. The *Ph. Eur.* criteria have since replaced the earlier criteria in the *BP* 1993.

12.2.6   Overall limitations

Compendial criteria by definition have to be applicable to the marketed product as available to the control analyst. Preservative efficacy therefore has to be based solely on the performance of the preserved formulation rather than on a comparative evaluation of an unpreserved and preserved formulation, which provides a more discerning method of assessment. Reproducibility is dependent *inter alia* on standardization of growth medium and inoculum preparation, by use of freshly prepared inocula. Cessation of use of stored inocula, as permitted in *USP* is necessary for harmonization of requirements.

Furthermore, use of a single microbial challenge with a few selected species cannot simulate the repeated in-use situation. Rechallenge of the preserved formulation is a more realistic test for simulation of the mode and frequency of use, where sequential challenges are undertaken at intervals of 2 or 7 days. The spectrum of the microbial challenge, however, can be improved by inclusion of strains isolated from contaminated formulations. These are invariably Gram-negative bacilli.

Preservative efficacy must be equated with the microbial purity requirement of the formulation to exclude pathogenic species. *P. cepacia*, which is prevalent in purified water and can also proliferate in marginally preserved formulation and diluents, is an opportunist pathogen in nosocomial infection (Bassett *et al.* 1970). Inclusion of this, or any species, when of importance for control of the microbial purity of the formulation is always advisable. This ensures that the preservative efficacy and microbial purity are investigated over the shelf-life of the formulation.

12.3   Cosmetics and toiletries

12.3.1   *Applicability of guidelines*

Guidelines for preservation of cosmetic and toiletry formulations produced in 1970 by the Toilet Goods Association (TGA) (Anon. 1970a) and the Society for Cosmetic Chemists (SCC) (Anon. 1970b) are less well defined than those applied to pharmaceuticals. However, a guideline produced by the Cosmetic, Toiletry and Perfumery Association (Anon. 1990b) recommends use of the *BP* 1988 challenge test for aqueous eye cosmetics with the interpretation criteria for topical preparations as stated in the *BP* 1988 (see Table 12.4).

The challenge test in the TGA is similar to that already described, with the unpreserved formulation included as a control. Formulations are examined at intervals for a minimum of 28 days as befits their nature and intended usage; therefore, some are examined for longer and some repeatedly rechallenged. No recommendations for adequacy of preservation are given, but a formulation with a relatively high count of the challenge species after 7 days may be considered inadequately preserved. A predictive judgement can be made after 14 days, with a final judgement at 28 days. Many formulations, however, would appear inadequately preserved if intended to exhibit similar fungicidal and bactericidal activity after 7 days, owing to the non-attainability of a 10^2 reduction of fungi in 7 days (see section 12.2.4).

12.3.2   *Tests with mixed cultures*

Procedures using mixed-culture challenges are described by the SCC for shampoos, creams, lotions and eye cosmetics. With shampoos, bacteria are selected to include *Pseudomonas* species isolated from spoilt shampoo or detergent preparations. These should be capable of vigorous growth and are cultured on media containing 5% unpreserved shampoo. Samples of the preserved and unpreserved formulation are inoculated, stored at 22°C and 30°C and bacterial counts estimated at intervals of 7 days for 28 days, or alternatively more frequently after 1 h, 6 h and 24 h and then daily for 1 week. The formulation is rechallenged where an immediate decrease to zero is observed and maintained for 2 consecutive weeks. The preserved formulation should have a self-sterilizing activity in 1–7 days, which is maintained for 28 days, against a count of 10^6 in the unpreserved formulation after 7 days. A slower bactericidal activity is acceptable if, for reasons of preservative stability a more rapidly acting preservative cannot be used.

Separate mixed cultures of bacteria, yeasts and moulds are used with preserved and unpreserved creams and lotions to include strains isolated from spoilt preparations and those obtainable from culture collections as shown in Table 12.9.

Samples are stored at 35–37°C for bacteria, 32°C for yeast and 22–25°C for mould rather than at 20–25°C as appropriate for determination of preservative efficacy. Bacteria and yeasts are examined by plate count over 2–3 months and a drastically reduced count or demonstration of no recovery, on three successive samples is regarded as indicative of good protection. Absence of mould growth is established visually by scattering a dry-spore inoculum of dark pigmented species over the surface of the cream and incubati‹ a moist atmosphere for 6 months. The method is incapable of direct differenti between fungicidal and fungistatic activity therefore and the guidelines for cos‹ creams and lotions are much less demanding than those of the TGA.

Table 12.9   Cultures recommended by the Society for Cosmetic Chemists (Anon. 1970b) for product challenge obtained from spoilt preparations or culture collections

Bacteria and yeasts	Moulds
Staphylococcus aureus	*Aspergillus niger*
Streptococcus faecalis	*Penicillium*
Pseudomonas aeruginosa	*Cladosporium*
Pseudomonas fluorescens	*Alternaria*
Escherichia coli	*Fusarium*
Klebsiella	*Rhizopus*
Proteus	*Phoma*
Candida albicans	*Trichoderma*
Saccharomyces cerevisiae	*Verticillium*
	Mucor

12.3.3   CTPA guideline for eye cosmetics

This guideline (Anon. 1990b) emphasizes the function of the preservative in consumer protection and prevention of spoilage during normal use, not as a replacement for good production hygiene. Preservative efficacy, with assessment based on the *BP* 1988 challenge test, should be ensured over the shelf-life by rechallenging stored product at intervals. Strains concomitant with the flora of raw materials, particularly water, the manufacturing environment and those isolated from contaminated aqueous eye cosmetics should be included. Sequential challenge of samples with inocula should be considered to simulate in-use conditions. Wet application of black mascara repeatedly with water, or saliva, is contra-indicated to prevent the microbial hazard inherent in this practice.

12.4   Conclusions

The imprecision of the TGA and SCC recommendations illustrate the difficulties in assigning criteria for preservation of formulations with a wide spectrum of use which contain preservatives of varying cidal and static activity. Realistic recommendations for preservation of pharmaceuticals, cosmetics and toiletries can only be gradually evolved by empirical evaluation to establish universally acceptable criteria. These must be capable of delineating between adequately and inadequately preserved preparations. Standardized and agreed test procedures are necessary for harmonization, but their modification is inevitable as more appropriate species are found for demonstration of preservative efficacy and the single challenge possibly replaced by a sequential challenge. Agreement, however, has first to be reached on a rate of kill and on stasis, which are applicable to preservatives with differing specific activities, to ensure adequate bacterial and mycotic preservation. This has to be achievable for a range of formulations with commonly used preservatives, where the efficacy may be enhanced, or decreased, by the constituents of the formulation and the container.

References

ANON. (1970a) A guideline for the determination of adequacy of preservation of cosmetics and toiletry formulations. *TGA Cosmet. J.* No. 2, 20–23.

——— (1970b) Hygienic manufacture and preservation of toiletries and cosmetics. *J. Soc. Cosmet. Chem.* **21** Appendix B 754–778.

——— (1980) Efficacy of antimicrobial preservatives in pharmaceutical products. In: *British Pharmacopoeia*. HMSO, London, Appendix XVIC, A192–A194.

——— (1988) Efficacy of antimicrobial preservatives in pharmaceutical products. In: *British Pharmacopoeia, Addendum*. HMSO, London, Appendix XVIC, A200–A203.

——— (1990a) Antimicrobial preservatives – effectiveness. In: *United States Pharmacopoeia XXII* US Pharmacopoeial Convention, Rockville, MD, 1478–1479.

——— (1990b) Microbial quality management CTPA limits and guidelines. Cosmetic, Toiletry and Perfumery Association 1990, Sect. B, 8; Sect. C 28–29.

——— (1993a) Efficacy of antimicrobial preservation. In: *British Pharmacopoeia*. HMSO, London, Appendix XV1C, A191–A192.

——— (1993b) Efficacy of antimicrobial preservatives. In: *European Pharmacopoeia* VIII 14, Fascicule 16.

——— (1994) In: *European Pharmacopoeia* VIII 14, Fascicule 17.

——— (1995a) Antimicrobial preservatives—effectiveness. In: *United States Pharmacopoeia XIII*. US Pharmacopoeial Convention, Rockville, MD, 1681.

——— (1995b) Antimicrobial preservatives—effectiveness (51) *Pharmacopoeial Forum*, **21**, 1040–1046.

——— (1995c) Efficacy of antimicrobial preservation. In: *British Pharmacopoeia (1993) Addendum*. HMSO, London. Appendix XVIIF, A407.

BASSETT, D. C. J., STOKES, K. J., and THOMAS, W. R. G. (1970) Wound infection with *Pseudomonas multivorans*: a water-borne contaminant of disinfectant solutions. *Lancet* **i** 1188–1191.

DAVISON, A. L., HOOPER, W. L., SPOONER, D. F., FARWELL, J. A., and BAIRD, R. M. (1991) The validity of the criteria of pharmacopoeial preservative efficacy tests – a pilot study. *Pharm. J.* **246** 555–557.

KURUP *et al.* reference omitted.

SPOONER, D. F., and DAVISON, A. L. (1993) The validity of the criteria for pharmacopoeial antimicrobial preservative efficacy tests. *Pharm. J.* **251** 602–605.

SYKES, G. (1958) The basis for sufficient of a suitable bacteriostatic in injections. *J. Pharm. Pharmacol.* **10** 40T–45T.

13

Challenge tests and their predictive ability

R. E. LEAK

Glaxo Research and Development Limited, Ware, Hertfordshire SG12 0DJ

C. MORRIS AND R. LEECH

Unilever Research, Quarry Road East, Bebington, Wirral, Merseyside L63 3JW

Summary

(1) Challenge tests which may be used during product development to give assurance of preservative efficacy are discussed.

(2) Results of a challenge test should indicate the degree of susceptibility of the formulation to microbial contamination.

(3) Two alternative methods of challenge testing are described which compare laboratory assessment of preservative activity with preservative performance during manufacture.

13.1   Introduction

A variety of challenge tests for the determination of preservative efficacy in pharmaceutical, cosmetic or toiletry products have been described, including those given in the *British Pharmacopoeia* (*BP*) (Anon. 1993a), the *European Pharmacopoeia* (*Ph. Eur.*) (Anon. 1994), the *United States Pharmacopoeia* (*USP*) XXIII (Anon. 1995) and the guidelines described by the Preservative Subcommittee of the Cosmetics, Toiletry and Fragrance Association (CTFA) (Halleck 1970) the Cosmetic, Toiletry and Perfumery Association (CTPA) (Anon. 1986, 1990) and the Society of Cosmetic Chemists (SCC) (Anon. 1970). The tests provide methods for assessing preservative efficiency in different product types and have been discussed in chapter 12. The harmonization of pharmacopoeial test methods has been recommended (Seyfarth and Lechner 1990, Geldsetzer 1990) and some proposals published (Anon. 1993b). The tests described in the *BP* and the *USP* present standards of preservative activity which should be attained by the preserved product, although those of the *BP* and *EP* are not intended as mandatory. These standards are defined with reference to the intended use of the product and provide a basis for the determination of the adequacy of the preservative system in a given product type. Although the compendial tests were designed for use with pharmaceutical products, they are often applied as a guideline in non-pharmaceutical products and the *USP* test is frequently applied to pharmaceutical product types outside its intended product range. The *BP* and *USP* tests have their limitations some of which have been discussed in chapter 12, section 12.2.5.

The objective of a challenge test is to determine the susceptibility of a product to microbial contamination. The *USP* and *BP* tests are not intended to mimic the conditions which may lead to contamination during manufacture and use and may not predict performance of preservative systems in practice (Cowen and Steiger 1976, Moore 1978, Ali-Hiti and Gilbert 1980). There is, nevertheless, evidence of correlation between preservative performance as measured using the *BP* challenge test and the magnitude of microbial contamination after use of some eye-drop products (Davision *et al.* 1991). Geis (1988), Sabourin (1990) and Singh (1987) have all reviewed the evaluation methods for cosmetic preservatives and contributed practical experiences from their own data on challenge testing. Some factors which contribute to the difference between performance of a preservative in the laboratory with its performance in practice are given below.

13.1.1   *The choice of test organisms and growth conditions*

The death rate of organisms has been shown to be affected by the growth medium and inoculum preparation procedure (Orth *et al.* 1989). Further, test organisms from culture collections grown on artificial media do not represent the behaviour of naturally occurring strains (Moore and Taylor 1976, Yablonski 1972, Leak 1983). The use of isolates from contaminated products, raw materials or the manufacturing environment is recognized in the preservative tests described in the *BP* and the guidelines published by CTFA and the CTPA. Such organisms are generally believed to be more aggressive than laboratory strains (Smith 1977, Cowen and Steiger 1976) and are more representative of the types of organism that a product will encounter during manufacture and use. One survey of challenge test data on 10,000 products indicated the difficulties involved in selecting test organisms (Spooner and Corbett 1985). Of the 8.5% challenge test failures, 55% were against only one organism. Of these single-organism failures, 75% were due to fungi, 15% to Gram-negative organisms, about 5% to yeasts and about 5% to Gram-positive cocci. From this work, Spooner and Corbett concluded that no single organism or group of organisms could be used to indicate whether a product is preserved against a wide range of bacteria, yeasts and moulds. The age of cells also affects resistance (Moore 1978) and the use of uniformly young cells, as described in the pharmacopoeial tests, does not simulate the conditions of cells contaminating a product in practice (Leitz, 1972).

13.1.2   *The number of challenges*

Laboratory tests based on a single microbial challenge do not reflect the practical circumstances of products which are subjected to a number of microbial challenges during use. Consequently, tests using multiple challenges were introduced (described in section 13.2.1).

13.1.3 Test length

The test length should be long enough to allow slow-growing organisms to emerge and on this basis a test period of 28 days may be too short (Allwood 1986). Regrowth of challenge organisms has been shown to occur outside this period (Leak 1983). A different view was taken by Spooner and Corbett (1985) who believed that the additional information obtained from extending tests beyond one week was not sufficient to justify the work involved.

A number of challenge tests are described in the following sections which can be used to supplement the information on preservative effectiveness obtained from the use of compendial tests. These tests are of particular use during product development and those described in the latter part of the chapter make a comparison of preservative effectiveness measured in the laboratory with that experienced during production.

13.2 Challenge tests for determining preservative effectiveness

13.2.1 Single-inoculation, repeat-inoculation and capacity tests

Preservative efficacy tests described in the *BP* and *USP* (discussed in chapter 12) are single-inoculation tests, i.e. the product is challenged with one inoculation of test organisms. Using the general method described in the *BP* or *USP*, testing can be performed using isolates relating to the manufacture or use of the product or from contaminated products, modifying the sample times and test length as required to give an appropriately detailed profile of the organism's loss of viability. Repeat-inoculation tests, in which the product is subjected to a series of inoculations of the challenge organism, were originally designed to simulate in-use conditions for certain types of product, e.g. topical pharmaceuticals and cosmetics, and were first recognized in the tests described by the CTFA (Halleck 1970) and the SCC (Anon. 1970). A wide variety of repeat-inoculation tests have been described in the literature (Yablonski 1972, Bruch 1975, Cowen and Steiger 1976, Londstrom 1986), differing in the number and frequency of inoculations. A study on the use of rechallenge in cosmetics carried out by the CTFA concluded that these tests can be effective in detecting marginally preserved systems (Anon. 1981). A successful reduction in the number of organisms following each repeat inoculation may indicate that the preservative system will demonstrate good efficacy in practice. This hypothesis will be considered in more detail later in this chapter (see section 13.3.2).

Tests in which the product is repeatedly challenged until the preservative system fails are called capacity tests. Barnes and Denton (1969) reinoculated test samples every 48 h with approximately 10^7 organisms g^{-1} or ml^{-1} for up to 15 inoculations. Samples were examined for surviving organisms prior to each reinoculation to determine the number of inoculations preceding the detection of survivors at three consecutive test points. One disadvantage of repeat-inoculation and capacity tests is that the product is progressively diluted with each inoculation. Another type of capacity test which does not dilute the product was described by Olsen (1967). It is a single-inoculation test, in which the test organism is inoculated into samples of preserved product using a range of inoculation levels, e.g. from 10^5 cfu ml^{-1} to 10^9 cfu ml^{-1}. The higher the inoculum that a test system can destroy, the greater the preservative activity and, by implication, the greater the subsequent product integrity. A further disadvantage of repeat-inoculation and capacity tests is that a high organic load is introduced into the product which may interact with some preservatives, reducing their activity (Flawn *et al.* 1973, Spooner and Crowshaw 1981).

13.2.2 Determination of preservative efficacy using D values

Preservative activity can be measured by assessing the initial rate of kill of the test organism. This can be achieved by determining the D value, the time required for 90% reduction in the population (the decimal reduction time). The preserved product is inoculated with a test organism and the number of viable cells determined at intervals over a short period, e.g. 6 h, 24 h or 48 h. From these data, the D value can be calculated by plotting log number of surviving organisms per gram as a function of the time after inoculation into the product (Orth 1980). Using this method, Orth (1980) calculated the D values of three concentrations of a given preservative in a product and determined the concentration of preservative required to give a specific D value. Orth (1981) suggested that D values of 4 h or less for pathogens and 28 h or less for non-pathogenic bacteria, yeasts and moulds (resulting in a destruction of 10^6 organisms g^{-1} in 24 h and 7 days, respectively) should give satisfactory preservation in cosmetics intended for use around the eyes and for baby products. D values can be used for evaluating different preservatives using a range of test organisms (Akers *et al.* 1984). For preservatives which produce a rapid initial reduction in microbial populations followed by a gradual subsequent loss of viability, Akers *et al.* (1984) suggested a quadratic expression for calculation of the D value which gave a more representative estimation of the destructive action of the preservative system than the linear model of Orth (1980). Orth and Milstein (1989) concluded from their work with D values and linear regression that preservative efficacy testing depended on several factors; reliability, time available, use of statistical controls, the need for rechallenge testing, the organisms used, the product being tested and the acceptance criteria. Testing the efficacy of the actual preservative was merely one part of the process of evaluating the preservative capacity of a product. Subsequently, Orth (1991) showed that products which complied with the standards of the *USP* and CTFA but which failed to meet acceptance criteria based on D values, occasionally failed to prevent growth of micro-organisms.

The use of D values has the advantage of providing data in 2–7 days and, therefore, allows a faster determination of efficacy than repeat-inoculation and capacity tests. Orth and Brueggen (1982) compared the D values obtained from repeat-inoculation challenges. Test systems were inoculated at intervals of 24 h with either *Pseudomonas aeruginosa* or *Staphylococcus aureus* and the D value determined for each inoculation. The total number of organism inoculated from 10 sequential challenges was calculated and this number was added to a fresh sample. The D values obtained from this test and from those after one challenge and 10 challenges in the previous tests were found to be the same. Orth and Brueggen concluded that little information could be gained from repeat challenges except to indicate when the system fails.

One of the problems of using D values for determining preservative efficacy is that challenge organisms can show an apparent complete loss of viability in the early stages of a challenge test (e.g. 1 h and 6 h after inoculation) but show regrowth some weeks later (Leak 1983). The relationship between initial rate of kill and subsequent regrowth has yet to be established. Spooner and Corbett (1985) observed that the majority of formulations which allowed survivors over a test period of four weeks showed viable organisms one week after challenge. Although the test period beyond one week detected a small proportion of formulations which allowed regrowth of survivors, the additional information from the four week test could hardly be justified.

13.2.3 *Mixed-culture challenge*

As already discussed, one factor which can significantly influence the outcome of a preservative test is the challenge organism and therefore testing frequently includes additional or alternative organisms to laboratory strains. In order to include a wide range of organisms but to minimize the number of tests to be performed, mixed-culture challenges rather than single-culture challenges have been advocated. Mixed-culture challenges also indicate the possible interaction between organisms that could occur in practice, which itself may influence preservative effectiveness (Moore 1978). In general, mixed-culture challenges consist of pools of Gram-positive organisms, Gram-negative organisms, yeasts or moulds. For example, Cowen and Steiger (1976) used a total of 11 organisms and Spooner and Crowshaw (1981) 41 organisms in this way. However, the use of such challenges has been questioned because mixed-culture challenges have been observed to indicate adequate preservation where single-culture challenges indicate inadequacy (Tagliapietra 1978). In particular, it has been reported that the mixed-culture challenge was less able to detect inadequate preservation against fungi (Yablonski 1972). Mixed cultures are not allow in some tests (Seyfarth 1993).

13.2.4 *Tests simulating in-use conditions*

In addition to the repeat inoculation tests described in section 13.2.1 a number of challenge tests have been described with reference to the determination of preservative activity in use.

Brannan *et al.* (1985) carried out *in vitro* 28-day challenge tests on four dilutions of shampoo and skin lotions using organisms isolated from contaminated products. Products were then classified as poorly preserved, marginally preserved or well preserved and consumer testing was carried out on each category to determine whether the challenge test had correctly predicted the risk of consumer contamination. Products classified as poorly preserved returned 46% to 90% contaminated after use, marginally preserved returned 0% to 21% contaminated after use and well preserved products, as classified by the *in vitro* challenge test, returned no contaminated products after use.

Ahearn *et al.* (1978) described a laboratory test of mascaras involving inoculation of the upper face of a 0.2 µm membrane filter positioned on the surface of the mascara which gave results comparable with those obtained from in-use testing of the mascaras. Orth *et al.* (1992) extended their work on *D* values to evaluate product preservation in relation to packaging and consumer use. Lindstrom and Hawthorne (1986) also studied the use of consumer tests to validate the microbiological integrity of cosmetic products.

The CTFA carried out a survey of challenge test methods used by cosmetic manufacturers (Curry 1990) and showed that 75% of the responding companies used *in vitro* and in-use testing of mascaras and 43% tested creams and lotions. Forty-six out of 53 companies correlated the results of in-use and *in vitro* testing and concluded that *in vitro* tests were predictive of in-use panel testing.

13.3 Challenge tests and preservative efficacy during manufacture and use

The objective of challenge tests is to predict the performance of the preservative during manufacture and use. Predictive tests can only be developed by comparison of the results of laboratory tests with experience of manufacture of products and with results of consumer-use tests.

The remaining part of this chapter describes two approaches which were adopted for the development of preservative tests; these attempted to predict product susceptibility to microbial attack. Section 13.3.1 presents an approach to preservative testing of pharmaceuticals which describes the development of a laboratory test and comparison of test results with manufacturing experience of a product, described in detail by Leak (1983). Section 13.3.2 describes a different approach involving a repeat-inoculation test which was developed for application to cosmetics and toiletries and where the results were compared with preservative performance of manufactured products.

13.3.1 *A laboratory test procedure for determining preservative efficacy during manufacture*

13.3.1.1 *Introduction*

Any potential product contaminants arising during manufacture will originate from the raw materials, from the environment (including the equipment) or from the operators. In other words, they will be wild strains entering the product via their natural habitat. The source of the test organism and its cultivation are likely to be important factors in the determined preservative efficacy. Strains isolated from contaminated products, water or the environment show a reduction in aggressive behaviour after subculture onto laboratory media (Gilbert *et al.* 1980, Carson *et al.* 1973). In a series of laboratory experiments (Leak 1983), the effect on resistance of a product contaminant cultivated for challenge testing in either its 'natural' habitat or on laboratory media was investigated, as shown in Fig. 13.1. The test organism, *Enterobacter cloacae*, contaminated an unpreserved sterile product, during manufacture. When cultivated in the sterile product and inoculated directly into the chlorhexidine diacetate 0.002% w/v solution the organism regrew from low levels of survivors but, after one subculture on tryptone soya agar, the organism showed a rapid loss of viability in the chlorhexidine solution. A laboratory strain of *E. cloacae*, grown on tryptone soya agar, also lost viability in the test system. Successive subcultures of the laboratory strain in the sterile product did not produce an organism that could reproduce the behaviour of the product strain (Fig. 13.2). Further experiments showed that other wild strains of *E. cloacae* (originally product contaminants, but maintained on tryptone soya agar) produced regrowth in the chlorhexidine system when grown in the sterile product. These results suggested that a stringent challenge of a test system was achieved by using a wild strain and by carrying over with the inoculum nutrients to which the test organism had adapted. This was supported by comparing the response of challenge organisms grown in the unpreserved environment (in this case water) to which preservative was added, with a tryptone soya agar culture of the organism inoculated into the preserved environment (Table 13.1). In this experiment, chlorhexidine diacetate (0.001% w/v) in sterilized distilled water was challenged using five strains of *Pseudomonas aeruginosa*. The organisms were grown in the sterilized water to

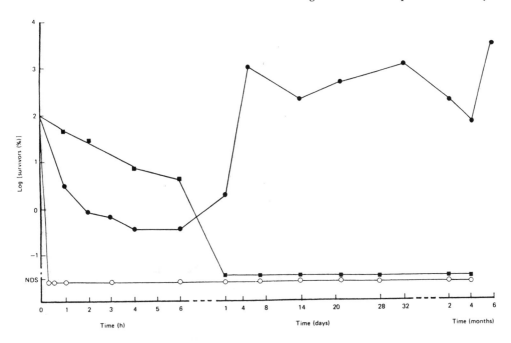

Figure 13.1 The sensitivity of *Enterobacter cloacae*, a contaminant of an unpreserved product (●, grown in unpreserved product; ■, grown on tryptone soya agar) and of a laboratory strain, NCTC 5920 (o, grown on tryptone soya agar) to chlorhexidine 0.002% w/v containing 1% v/v unpreserved product. The 1% unpreserved product was either added to the chlorhexidine solution with the inoculum, when the organism was grown in unpreserved product, or added immediately prior to inoculation, for test organisms growth on tryptone soya agar. NDS indicates no detectable survivors

10^5–10^6 cfu ml^{-1} to which was added a concentrate of chlorhexidine (0.5% w/v) to give a 0.001% w/v solution. The organisms were also grown on tryptone soya agar and inoculated into chlorhexidine (0.001% w/v) to give 10^6 cfu ml^{-1}. In all tests the organisms showed a rapid loss of viability but varying abilities to regrow. For four out of five organisms, regrowth occurred more frequently when the organism was challenged in its growth habitat (water).

This approach to challenge testing was adopted for testing an oral pharmaceutical product with a known microbiological history. The results of these tests were compared with those from compendial tests in relation to their ability to establish the adequacy of the preservative systems. The product tested was preserved at that time using a combination of methyl and propyl parabens in a ratio of 2:1 to give a total of 0.46% w/v (i.e. at a concentration above the solubility of these preservatives in water). However, a lower but nonetheless effective concentration of parabens was also being sought at that time.

13.3.1.2 Laboratory tests

(1) **Results of the *USP* preservative effectiveness test** – Performance of the *USP* test (Anon. 1985) indicated that at concentrations of both 0.46% and 0.15% w/v parabens was an effective preservative system, less than 10 cfu ml^{-1} of all challenge organisms being detectable at 1 week and thereafter for 4 weeks.

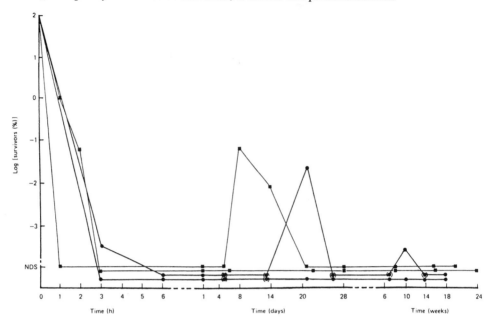

Figure 13.2   The sensitivity of a laboratory strain of *Enterobacter cloacae*, NCTC 5920, grown in an unpreserved product to chlorhexidine 0.002% w/v containing 1% unpreserved product added via the inoculum: ■, three subcultures in unpreserved product prior to testing; ●, 10 subcultures in unpreserved product prior to testing; (●), no survivors by surface viable count, but growth in the recovery medium. NDS indicates no detectable survivors.

Table 13.1   The effect of growth environment on the response of five strains of *Pseudomonas aeruginosa* to chlorhexidine diacetate (0.001% w/v) in water

	Number of challenge tests in which regrowth occurred (out of three)	
Strains of *Pseudomonas aeruginosa*[a]	Grown on tryptone soya agar and inoculated into chlorhexidine diacetate (0.001% w/v) at 10^6 cfu ml^{-1}	Grown in water to 10^5–10^6 cfu ml^{-1}; chlorhexidine diacetate solution added to give a chlorhexidine diacetate 0.001% w/v solution
P_1	1	3
P_3	1	3
P_3	2	2
P_4	1	2
P_L	0	3

[a] P_1–P_4, wild strains; P_L, NCTC 6750.

(2)   **Additional laboratory challenge tests** – Further challenge tests were carried out using the approach described in section 13.3.1.1 on product containing 0%, 0.05%, 0.15%, 0.2% and 0.46% w/v parabens. In order to relate the tests to the manufacture of the product, test organisms were selected from a range of organisms isolated from the raw materials, from the production environment, from samples of product part way through the manufacturing process and from unpreserved product.

P. aeruginosa from the manufacturing environment and *E. cloacae* from the unpreserved product were chosen as representative isolates. The *P. aeruginosa* was demonstrated to grow from 10^3 cfu ml^{-1} to 10^6 cfu ml^{-1} in the unpreserved product.

Two methods for cultivating challenge organisms were adopted for challenge testing, avoiding cultivation on laboratory media. First, test organisms were grown in unpreserved product and the adapted organisms inoculated directly into the preserved product. Secondly, since a notorious source of product contamination is the manufacturing water, test organisms were grown in manufacturing water and the water then used to prepare the preserved product to simulate contamination arising in this way. The organisms were growth through three sequential subcultures in sterilized production water or in unpreserved product for test. The inoculum size used in the test was based on the manufacturer's microbial limits of 10^2 cfu ml^{-1} for manufacturing water and for the finished product. This was raised by a factor of 10 to 10^3 cfu ml^{-1} to allow a margin of safety.

P. aeruginosa and *E. cloacae* grew to levels of about 10^5 cfu ml^{-1} in sterilized manufacturing water. The water was diluted with sterilized manufacturing water to give 10^3 cfu ml^{-1}. Using this water, 100-ml quantities of product containing parabens concentrations of 0%, 0.05%, 0.15%, 0.2% and 0.46% w/v in duplicate were prepared according to the manufacturing method.

P. aeruginosa and *E. cloacae* grew to approximately 10^6 cfu ml^{-1} in unpreserved product. Volumes of 50 ml of product containing 0%, 0.05%, 0.15%, 0.2% and 0.46% w/v parabens were inoculated in duplicate with 50 μl of unpreserved product containing the test organism giving 10^3 cfu ml^{-1} in the final test system. All test systems were stored at 25°C and sampled at intervals to estimate the numbers of survivors. The results of these tests are given in Figs 13.3 and 13.4 and described below.

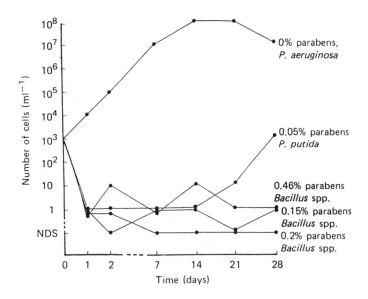

Figure 13.3 Challenge tests of an oral pharmaceutical product containing a range of concentrations of parabens. The test organism *Pseudomonas aeruginosa* was grown for testing in unpreserved product. The organisms isolated from test systems were *Pseudomonas aeruginosa*, *Pseudomonas putida* (source unknown) and *Bacillus* spp. (from one of the raw materials). Similar results were obtained in duplicate tests. NDS indicates no detectable survivors.

207

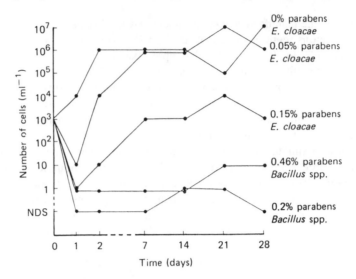

Figure 13.4 Challenge tests of an oral pharmaceutical product containing a range of concentrations of parabens. The test organism *Enterobacter cloacae* was grown for testing in unpreserved product. The organisms isolated from test systems were *Enterobacter cloacae* and *Bacillus* spp. (from one of the raw materials). Similar results were obtained in duplicate tests. NDS indicates no detectable survivors

P. aeruginosa and *E. cloacae* were unable to survive in any of the test systems when cultivated in the manufacturing water used to prepare the preserved and unpreserved (control) test products (testing continued for up to 4 months). The importance of inoculating the unpreserved product is indicated here, since without this it may have been incorrectly concluded that all concentrations of parabens were capable of preventing survival of the test organism. In contrast, when *P. aeruginosa* was grown in unpreserved product, it showed survival and growth in the unpreserved control test product but lost viability in all the test products containing parabens (no regrowth detected during a test period of 2 months) (see Fig. 13.3).

E. cloacae, cultivated in unpreserved product, also showed survival and growth in the unpreserved control test product but, whereas preservation was achieved with 0.2% and 0.46% parabens, 0.05% and 0.15% parabens were shown to be inadequate, allowing regrowth of the test organism to about 10^3 survivors ml^{-1} and 10^6 survivors ml^{-1}, respectively (Fig. 13.4). No regrowth was observed in the products containing 0.2% or 0.46% parabens in further testing at 2 months. The results are in contrast with those obtained from *USP* tests described in (1) above, which indicated that 0.15% w/v parabens provided good preservation of the product. However, results from all the laboratory tests ((1) and (2) above) indicated that product containing 0.46% w/v parabens was well preserved.

13.3.1.3 *Production experience*

The manufacturer had considerable experience of the product containing 0.46 w/v parabens, which compared favourably with the results of the laboratory tests. Of 229 batches of product, 161 had a total viable mesophilic aerobic count of less than

10 cfu ml^{-1}, 67 between less than 10 cfu ml^{-1} and 10 cfu ml^{-1} and one of 151 cfu ml^{-1}. The organisms were *Bacillus* spp., assumed to have originated from some of the raw materials. On one occasion, a batch of product was found to contain a strain of *E. cloacae* which caused degradation of the parabens esters. Attempts to contaminate fresh product (0.46% parabens) with product containing the contaminant were unsuccessful. Perform-ance of the *USP* challenge test using an isolate of the product contaminant demonstrated less than 10 cfu ml^{-1} of the organism at one week and thereafter for four weeks. However, the susceptibility of the product to *E. cloacae* contamination was predicted in laboratory tests in the isolation of this organism from batches of unpreserva product prepared in the laboratory.

On the basis of the *USP* test results (Anon. 1985), the manufacturer selected the lower concentration of 0.15% w/v parabens for preserving the product but in practice experienced some problems during the early stages of manufacture of this formulation. This experience was consistent with the conclusions drawn from laboratory challenge tests on product containing 0.15% w/v using *E. cloacae* cultivated in unpreserved product (Fig. 13.4) discussed in (2) above.

These results suggest that using a test strain isolated from unpreserved product and grown for challenge testing in unpreserved product provides a means of detecting marginally preserved systems. A similar type of test was described by Spooner and Crowshaw (1981) where mixed cultures were inoculated into unpreserved product. After four weeks, these organisms were used in challenge tests and, if poor results were obtained, these products were monitored more frequently during production and filling.

13.3.2   A predictive repeat-inoculation test

13.3.2.1   Introduction

The objective of a challenge test is to determine the risk of susceptibility to microbial contamination. Rather than the pass or fail results associated with 'official' tests, a more realistic challenge test result would be a risk assessment based on a sliding scale from low to high risk. As a result, no formulation would fail a challenge test; it would just have a medium to high risk of preservative failure. Similarly, a formulation that passed a 'traditional' challenge test would have a low to medium risk of preservative failure. This section describes the development of a test method to provide these risk data.

13.3.2.2   Repeat challenge

One of the problems with traditional challenge tests is that small differences in the status of micro-organisms used in the test can make a large difference to the final challenge result (see section 13.3.1). If the micro-organisms are too aggressive to the formulation, then product will fail even though, by most other tests and experience, they would be considered adequately preserved. The use of repeat-challenge tests, where formulations are repeatedly challenged until microbial survivors are found, provides a means not only of assessing the aggressiveness of specific inocula but also of ranking various formula-tions in terms of their susceptibility to contamination. In one such test (see Table 13.2) using two different inocula (a laboratory strain and a product contaminant), it was ob-served that the test length varied depending on the aggressiveness of the inoculum, but not the ranking of the formulations. Thus, formulation A would be considered high risk

Table 13.2   Illustration of repeat-challenge-test results

	Number of repeat inoculations until microbial survival observed	
	Type culture (laboratory) strain	Product contaminant
Unpreserved control	10	1
Formulation A	30	5
Formulation B	165	32
Well-preserved control	150	30

whichever test inoculum was used. It was therefore decided to investigate the repeat challenge further and see whether it could be applied to cosmetic and toiletry products.

13.3.2.3   *Micro-organisms*

In any microbial bioassay the micro-organisms selected must be representative of those found *in vivo* and sufficiently aggressive to allow separation of test formulations. Obviously a weak microbial inoculum would mean that the test would run too long and an aggressive inoculum would not give a large enough spread of failure points to separate formulations. To this end, type cultures were found to be too weak whereas contaminants in formulations were too aggressive. It was finally decided to use Gram-negative isolates from contaminated formulations as the bacterial challenge. Yeasts and fungi were similarly isolated. The other factor found to be important in inoculum selection was formulation pH. The inocula selected for low- or medium-pH formulations were unsuitable for high-pH formulations and vice versa, i.e. the inoculum was too weak outside its own pH range. As a result, two groups of inocula were gathered: one for neutral- or low-pH formulations and one for high-pH formulations. Cocktails of micro-organisms were produced to cover variations in microbial behaviour. The use of cocktails reduced the number of samples processed for each test. Each cocktail was restricted to each microbial type, i.e. bacteria, yeast or fungi, and they were checked to ensure that there was no microbial antagonism occurring between the micro-organisms. It was decided to use a high inoculum to increase the challenge potential as recommended in many standard challenge test methods (Anon. 1993a, 1993b).

13.3.2.4   *Inoculum preparation*

Bacteria were isolated on tryptone soya agar (Oxoid) and incubated at 28°C for 3–5 days. Yeasts and moulds were similarly isolated on Sabouraud dextrose agar (Oxoid). These cultures were used with subculturing for 2–3 months, after which fresh micro-organisms were re-isolated from products. Bacterial and yeast inocula were prepared by transferring one colony into 10 ml tryptone soya broth (Oxoid) or Sabouraud dextrose broth, incubating at 28°C for 24 h and then transferring 2 ml into 200 ml of the respective broths. After 24 h at 28°C, micro-organisms were recovered by centrifugation (4,000 rev min^{-1} for 10 min), washed twice in sterile distilled water and finally resuspended in 100 ml sterile distilled water. This gave an inoculum concentration of approximately 10^9 cfu ml^{-1}. Mould spores were harvested from Petri dish culture using 5–10 ml of 1% Tween 80 per dish.

13.3.2.5   Experimental procedure

The inoculum size was 1 ml in 99 ml of test formulation, giving a test microbial concentration of 10^6–10^7 cfu ml^{-1}. Formulations were inoculated every 24 h after a 1 ml sample had been withdrawn for total viable count. The test for each formulation was stopped when viable micro-organisms were isolated from the formulation on two consecutive days at greater than 10^2 cfu ml^{-1}. Formulations were incubated at 25°C.

13.3.2.6   Initial results

A range of formulations for one product type was tested and the number of inoculations to failure determined, as shown in Table 13.3. From these results the products were ranked from high risk (two inoculations to failure) to very low risk (50 inoculations to failure). As a comparison with this ranking, data were collected on the product risk as determined by factory experience and quality data. These data separated the formulations into three broad categories: high, medium and low risk. There was a very close agreement between the factory product ranking and the challenge results (Table 13.3). The only formulation that did not correlate was product I which by the challenge test appeared to be medium risk but by the quality data appeared to be low risk. There was insufficient quality data, however, to be sure of the factory ranking.

If the repeat challenge test indicates that a product is high risk with regard to microbial contamination it does not mean that the formulation cannot be manufactured. It does, however, mean that the strictest control must be applied to the microbiological quality of the raw materials and their handling, that factory hygiene must be of the highest standard and that the product packaging should prevent in-use contamination.

13.3.2.7   Comparative tests

An alternative method was to use only a single inoculation of the test product but to use a range of dilutions to determine the preservative failure point. It was found that the end

Table 13.3   Comparison of repeat-challenge data with factory experience of formulations

Formulation	Challenge test			Factory or quality data		
	Number of inoculations to failure	Contamination risk estimated from challenge data	Formulation ranking (high to low risk)	Contamination risk	Formulation ranking	
A	2	High	A	High	A	
B	10	Medium	C	High	C	
C	2	High	E	High	E	
D	48	Low	G	Medium	G	
E	2	High	J	Medium	J	
F	24	Low	B	Medium	B	
G	7	Medium	I	Low	I	
H	22	Low	H	Low	H	
I	15	Medium	F	Low	F	
J	7	Medium	K	Low	K	
K	47	Low	D	Low	D	

points for many of the tested formulations occurred within a small range of dilutions, making it difficult to assess the relative risks of preservation failure. Overall the dilution test was rejected as a suitable challenge method for the following reasons.

(1) The results did not correlate with factory experience. This was probably due to the 'clustering' of results explained above.

(2) With the dilution challenge, micro-organisms were not exposed to the use concentration of the product as with the repeat challenge.

(3) The dilution challenge did not allow for any microbial adaptation to the formulation to occur, which may be an important factor for certain preservative systems (Kato *et al.* 1983).

13.3.2.8   Discriminative tests

Further investigations using the repeat-challenge method indicated that it would provide an excellent tool for the development chemist/formulator to assess the effect of increasing concentration of one preservative or the relative activity of preservatives in mixed preservative systems.

Table 13.4 shows the results of a challenge test on a shampoo formulation containing increasing levels of the same preservative at a constant pH and the effect of increasing pH at the highest concentration of preservative as tested. The products were challenged with a mixture of Gram-negative bacteria and with *Candida parapsilosis* (yeast) isolated from a contaminated shampoo. The test showed that increasing the preservative concentration improved the preservative capacity of the product against bacteria but not against yeasts. Increasing the pH value reduced the antibacterial activity but again did not noticeably affect the anti-yeast activity. It can be predicted that the bacteria were sensitive to both the preservative concentration and the pH of the formulation and that the yeast was relatively unaffected by this preservative or by low pH.

The repeat challenge test is therefore of value in investigating formulation susceptibility to a variety of micro-organisms under different chemical and physical preservation regimes.

13.3.2.9   The use of rapid methods for preservative efficacy testing

Rapid microbiological techniques have been developed in recent years as an alternative to the traditional total viable count procedures. The principles of several of these methods are reported in Chapter 14.

Table 13.4   Challenge-test results against different inocula

Product	Number of inoculations to failure		
	Bacteria	Yeast	pH
Shampoo + 0.3% preservative	20	6	5.4
Shampoo + 0.4% preservative	25	6	5.4
Shampoo + 0.5% preservative	35	6	5.4
Shampoo + 0.5% preservative	15	6	5.9
Shampoo + 0.5% preservative	7	5	6.4

The major advantage of techniques such as impedance, direct epifluorescence (DEFT-MEM) and ATP bioluminescence (ATP-B) lie in the more economic use of materials and the reduction in time for results to be obtained.

De Pasquale *et al.* (1985) used impedance techniques to assess the effectiveness of preservative combinations in hand lotion. Calibration curves were constructed for *E. coli*, *S. aureus* and *P. aeruginosa* to relate detection times to plate counts over a range of inocula, 10^3–10^7 cfu g^{-1}. Standardization of the protocol resulted in good correlation curves for the test organisms. Detection time could then be applied to the appropriate calibration curve to estimate the surviving population after exposure to different preservatives. De Pasquale *et al.* concluded that impedance techniques reduced sample preparation and processing times and provided an automated method of evaluating preservative activity in cosmetic products.

Connolly, Bloomfield and Denyer (1993a) carried out a comparative study of the use of impedance, DEFT-MEM and ATP-B techniques for rapid evaluation of the preservative efficacy of pharmaceuticals and cosmetics. A good correlation was obtained for untreated suspensions of *S. aureus*, *P. aeruginosa* and *Candida albicans* between rapid method responses and total viable counts using all three methods but only impedance gave a good correlation for *Aspergillus niger*. After treatment with chlorhexidine, good dose response curves were obtained for *S. aureus* and *Candida albicans* using impedance but ATP-B and DEFT-MEM methods underestimated the kill by an order of 1–6 log units.

Connolly, Bloomfield and Denyer (1993b) extended their work with impedance to look at the effects of methyl parabens and phenoxyethanol in phosphate buffered saline and in cetomacrogol cream. The work showed that correlations between total viable count and impedance measurements were significantly different for cells which had been exposed to preservatives than for untreated cells. For any given number of colony forming units the time to detection for preservative treated cells was extended. The group concluded, however, that impedance was a valid method for preservative testing, particularly for economically screening formulations against a wider range of preservatives and varying concentrations of use than could currently be handled by traditional methods because of the workload involved.

Undoubtedly the use of rapid methods for preservative efficacy testing is an area where considerable attention will be focused in the future.

13.4   Conclusion

The importance to the manufacturer of determination of preservative efficacy is clearly indicated by the variety of challenge tests which are in use and which are undergoing continual development. In particular, the development of challenge tests, the results of which can be interpreted with reference to an assessment of risk of product contamination, is paramount. The success of a predictive challenge test is judged by the degree of correlation between laboratory results and production experience. Both tests described above (see sections 13.3.1 and 13.3.2) demonstrate the ability to predict the susceptibility of products to contamination. The first approach, using a test strain isolated from and cultivated in unpreserved product, however, needs to be applied to further products to confirm its predictive ability. The repeat-challenge test appears to be capable of realistically assessing contamination risk and has the advantage that the test is not reliant on the aggressiveness of the test organisms. Both tests were compared with production experience of preservative efficacy but as yet this comparison has not been extended to experience in use.

References

AHEARN, D. G., SANGHVI, J., and HALLER, G. J. (1978) Mascara contamination: in-use and laboratory studies. *J. Soc. Cosmet. Chem.* **29** 127–131.

AKERS, M. J., BOAND, A. V., and BINKLEY, D. A. (1984) Preformulation method for parenteral preservative efficacy evaluation. *J. Pharm. Sci.* **73** 903–905.

AL-HITI, M. M., and GILBERT, P. (1980) Changes in preservative sensitivity for the *USP* antimicrobial agents effectiveness test micro-organisms. *J. Appl. Bacteriol.* **49** 119–126.

ALLWOOD, M. C. (1986) Preservative efficacy testing of pharmaceuticals. *Pharma. Int.* **7** 172–175.

ANON. (1970) The hygienic manufacture and preservation of toiletries and cosmetics. *J. Soc. Cosmet. Chem.* **21** 719–800.

(1981) A study of the use of rechallenge in preservation testing of cosmetics. *CTFA Cosmet. J.* **13** 19–22.

(1985) Antimicrobial agents – effectiveness. In: *United States Pharmacopoeia XXI.* US Pharmacopoeial Convention. Rockville, MD.

(1986) *CTPA recommended microbiological limits and guidelines to microbiological quality control.* Cosmetic, Toiletry and Perfumery Association, London.

(1990) Microbial quality management CTPA limits and guidelines. *Cosmetic, Toiletry and Perfumery Association*, Section B 8, Section C 28–29.

(1993) Efficacy of antimicrobial preservation. *British Pharmacopoeia*, HMSO, London, Appendix XVI C, A191–A192.

(1993b) Efficacy of antimicrobial preservation. *Pharmeuropa* **5** 355–358.

(1994) Efficacy of antimicrobial preservation. *European Pharmacopoeia*, Maisonneuve SA, France, VIII.14-1–VII.14-4.

(1995) Antimicrobial agents – effectiveness. *United States Pharmacopoeia XXIII.* US Pharmacopoeial Convention, Rockville, pp. 1478–1479.

BARNES, M., and DENTON, G. W. (1969) Capacity tests for the evaluation of preservatives in formulations. *Soap, Perfum. Cosmet.* **42** 729–733.

BRANNAN, D. K., DILLE, J. C., and KAUFMAN, D. J. (1985) Correlation of *in vitro* challenge testing with in-use consumer testing for cosmetic products. *J. Soc. Cosmet. Chem.* **36** 375–380.

BRUCH, M. K. (1975) The regulation of hydrophilic contact lenses by the Food and Drug Administration. In: Underkofler, L. A. (ed.), *Developments in industrial microbiology.* Pharmaceutical Press, New York, pp. 29–47.

CARSON, L. A., FAVERO, M. S., BOND, W. W., and PETERSEN, N. J. (1973) Morphological, biochemical and growth characteristics of *Pseudomonas cepacia* from distilled water. *Appl. Microbiol.* **25** 476–483.

CONNOLLY, P., BLOOMFIELD, S. F., and DENYER, S. P. (1993a) A study of the use of rapid methods for preservative efficacy testing of pharmaceuticals and cosmetics. *J. Appl. Bacteriol.* **75** 456–462.

(1993b) The use of impedance for preservative efficacy testing pharmaceutical and cosmetic products. *J. Appl. Bacteriol.*

COWEN, R. A., and STEIGER, B. (1976) Antimicrobial activity – a critical review of test methods of preservative efficiency. *J. Soc. Cosmet. Chem.*, **27** 467–481.

CURRY, A. (1990) CTFA survey: test methods companies use. *Cosmet. Toilet.* **105** 79–82.

DAVISON, A. L., HOOPER, W. L., SPOONER, D. F., FARWELL, J. A., and BAIRD, R. M. (1991) The validity of the criteria of pharmacopoeial preservative efficacy tests – a pilot study. *Pharm. J.* **246** 555–557.

DE PASQUALE, D., KAHN, P., and FIRSTENBERG-EDEN, R. (1985) Impedimetric determination of antimicrobial preservative efficacy: an automated challenge test. *Abst. Ann. Meet. Am. Soc. Microbiol.* **85.**

FLAWN, P. C., MALCOLM, S. A., and WOODROFFE, R. C. S. (1973) Assessment of the preservative capacity of shampoos. *J. Soc. Cosmet. Chem.* **24** 229–235.

GELDSETZER, K. (1990) Preservation: preservative efficacy test (PET) in national pharmacopoeias. *Dtsch. Apoth. Ztg. Dazea* **130** 502–507.

GEIS, P. A. (1988) Preservation of cosmetics and consumer products: rationale and application. *Dev. Ind. Micro.* **29** 305–315.

GILBERT, P., AL-HITI, M. M., and BEVERIDGE, E. G. (1980) Rational approach towards improving the reproducibility and predictive effectiveness of 'official' antimicrobial preservative challenge tests. *J. Pharm. Pharmacol.* **32** (suppl.) 14P.

HALLECK, F. E. (1970) A guideline for the determination of adequacy of preservation of cosmetic and toiletry formulations. *TGA Cosmet. J.* **2** 20–23.

KATO, N., MIYAWAKI, N., and SAKAZAWA, C. (1983) Formaldehyde dehydrogenase from formaldehyde resistant *Debaryomyces vanriji* FT-1 and *Pseudomonas putida* F61. *Agric. Biol. Chem.* **47** 415–416.

LEAK, R. E. (1983) *Some factors affecting the preservative testing of aqueous systems*, Ph.D. Thesis. University of London.

LEITZ, M. (1972) Critique of *USP* microbiological test. *Bull. Parenter. Drug Assoc.* **26** 212–216.

LINDSTROM, S. M. (1986) Consumer use testing: assurance of microbiological product safety. *Cosmet. Toilet.* **101** 71–73.

LINDSTROM, S. M., and HAWTHORNE, J. D. (1986) Validating the microbiological integrity of cosmetic products through consumer use testing. *J. Soc. Cosmet. Chem.* **37** 481–488.

MOORE, K. E. (1978) Evaluating preservative efficacy by challenge testing during the development stage of pharmaceutical products. *J. Appl. Bacteriol.* **44** Sxliii–Slv.

MOORE, K. E., and TAYLOR, J. E. (1976) Microbiological standards for nasal solutions. *J. Appl. Bacteriol.* **41** 379–387.

OLSEN, S. W. (1967) The application of microbiology to cosmetic testing. *J. Soc. Cosmet. Chem.* **18** 191–198.

ORTH, D. S. (1980) Establishing cosmetic preservative efficacy by use of *D*-values. *J. Soc. Cosmet. Chem.* **31** 165–172.

(1981) Principles of preservative efficacy testing. *Cosmet. Toilet.* **96** 43–52.

(1991) Standardizing preservative efficacy test data. *Cosmet. Toilet.* **106** 45–48.

ORTH, D. S., and BRUENGGEN, L. R. (1982) Preservative efficacy testing of cosmetic products. *Cosmet. Toilet.* **97** 61–65.

ORTH, D. S., and MILSTEIN, S. R. (1989) Rational development of preservative systems for cosmetic products. *Cosmet. Toilet.* **104** 91–103.

ORTH, D. S., LUTES, C. M., and SMITH, D. K. (1989) Effect of culture conditions and method of inoculum preparation on the kinetics of bacterial death during preservative efficacy testing. *J. Soc. Cosmet. Chem.* **40** 193–204.

ORTH, D. S., BARLOW, R. F., and GREGORY, L. A. (1992) The required *D*-value; evaluating product preservation in relation to packaging and consumer use/abuse. *Cosmet. Toilet.* **107** 39–43.

SABOURIN, J. R. (1990) Evaluation of preservatives for cosmetic products. *Drugs. Cosmet. Ind.* **147** 24–27, 64–65.

SEYFARTH, H. (1993) Test for efficacy of antimicrobial preservation. The regulations of the DAB 10 in comparison of the DAB 10 Addendum 1992. *Pharm. Ind.* **55** 387–391.

SEYFARTH, H., and LECHNER, U. (1990) Test for efficacy of preservation according to the regulations of DAB 9 (German Pharmacopoeia). *Pharm. Ind.* **52** 1129–1142.

SINGH, V. B. (1987) Practical experience in the preservation of cosmetic formulations. *Parfuem. Kosmet.* **68** 752–770.

SMITH, J. L. (1977) Product testing for preservation efficacy. *Cosmet. Toilet.* **92** 30–34.

SPOONER, D. F., and CORBETT, R. J. (1985) The rational use of preservatives in cosmetics and toiletries. *Symposium on Formulating Better Cosmetics*, University of Nottingham.

SPOONER, D. F., and CROWSHAW, B. (1981) Challenge testing – the laboratory evaluation of the preservation of pharmaceutical preparations. *J. Microbiol. Serol.* **47** 168–169.

TAGLIAPIETRA, L. (1978) Antimicrobial power of liquid and semi-solid pharmaceutical preparations. *Cosmet. Toilet.* **93** 23–26.

YABLONSKI, J. I. (1972) Fundamental concepts of preservation. *Bull. Parenter. Drug Assoc.* **26** 220–227.

14

New methodology for microbiological quality assurance

E. M. WATLING AND R. LEECH

Unilever Research, Quarry Road East, Bebington, Wirral, Merseyside L63 3JW

Summary

(1)   New methods for microbiological quality assurance (MQA) are now available as replacements for traditional plating techniques.

(2)   These new methods are faster than traditional techniques and most are more automated and labour saving.

(3)   New methods must be subjected to a development and validation phase before they can replace existing plating techniques.

14.1   Introduction

MQA is a vital function within the pharmaceutical, cosmetic and toiletries industries. It is required for a variety of functions from checking raw material quality, through monitoring hygienic processing, to checking final product quality. Two basic functions are linked with MQA: detection and enumeration of micro-organisms followed by identification of contaminants. This chapter will concentrate on the first of these functions where new technology is being investigated to replace existing methods. Identification methods using new technology are now fairly well established throughout the industry, many of these techniques being considerably faster than traditional biochemical testing (Homes *et al.* 1977, Nord *et al.* 1974, Griffiths and Phillips 1982, Palmieri *et al.* 1988). Some of the newer identification techniques will be briefly discussed at the end of the chapter.

Present detection and enumeration methodology is widely based on traditional microbiological techniques (Madden 1984, Hooper 1981). These techniques have been discussed earlier in chapter 3 and further details can be found in Baird (1990). Agar plate counts, whether using non-selective or selective media, are slow to perform, require skilled handling and interpretation and are labour intensive. Owing to the incubation delay for traditional techniques, MQA rarely acts as much more than a retrospective quality monitor. Raw materials have to be accepted before microbial counts are available, product is processed on equipment for which microbial loads are unknown and product is packed and in storage before microbial quality results are known. As a consequence industry needs faster, automated, less expensive methods for MQA and this chapter reviews the methods available and their applicability.

14.1.1 Microbial quality assurance requirements

The requirements for MQA in the various industries are summarized in Table 14.1. The ideal system for final product analysis would be continuous monitoring using in-line microbiological sensors. Biosensors currently available for this purpose are not sensitive enough for the industry (see later) although, provided that the biological significance of a chemical factor is known, then chemical sensors could have a role to play in MQA. For example, if a product is preserved by reduced water-activity, quality assurance could theoretically be monitored by water activity measurements rather than by plate counts. This sort of example is, unfortunately, not often encountered in these industries and biological monitoring is therefore usually adopted.

14.1.2 Rapid methods

There have been a variety of methods suggested to fulfil the requirements of the industries. These are listed in Table 14.2. They represent a wide variety of techniques and degrees of automation. Some methods are supported by a range of commercially available equipment whilst others have little or none. Capital cost for equipment varies considerably as does the method sensitivity and applicability. Much of the data that follow have been accumulated from monitoring of toiletries although applications for cosmetics and non-sterile pharmaceuticals are not too dissimilar. It is almost certain that all the methods discussed can be used for MQA in these industries providing that they are tailored and validated against the formulations and micro-organisms encountered by the respective industry. In addition, in the case of pharmaceuticals, any selected method must satisfy the regulatory authorities with regard to sensitivity and/or specificity.

14.1.3 Micro-organisms associated with contamination

Table 14.3 lists organisms which typically cause contamination problems in pharmaceuticals, toiletries and cosmetics. The list is divided into two groups: those that are mainly isolated and those that occur less frequency. Other organisms have also occasionally

Table 14.1 The requirements for microbial quality assurance of pharmaceuticals, cosmetics and toiletries

Pharmaceuticals	Cosmetics and toiletries
Total viable counts	Total viable counts
Absence of, or maximum number of, specific pathogens	
Challenge testing	Challenge testing
Sterility tests	

Table 14.2 Rapid methods

Method	Reference	Principle of detection
Electrical	Cady *et al.* (1978) Kell and Davey (1990)	Electrical changes in media due to microbial growth
ATP bioluminescence	Chappelle and Leving (1968) Nielsen and Van Dellan (1989)	Detection of microbial ATP using bioluminescent reagent
Direct epifluorescent filtration technique	Pettipher (1983) Newby (1991)	Microbial staining and counting on filters
Microcalorimetry	Forrest (1972)	Detection of microbial heat generation
Flow cytometry	Fung (1991) Kaprelyants and Kell (1992)	Microbial counting
Limulus amoebocyte lysate	Cohen (1979)	Gram-negative lipopolysaccharide detection
DNA/RNA probes	Tomkins *et al.* (1985) Fitter *et al.* (1992)	Microbial DNA/RNA hybridization
Enzyme monitoring and detection of electron transfer	Kroll and Rodrigues (1986) Kroll *et al.* (1989)	Detection of microbial enzymes, e.g. cytochrome oxidase
Enzyme-linked immunosorbent assay	Granstroem *et al.* (1984) Gonzalez *et al.* (1993)	Antigen–antibody complexes
Immunomagnetic separation	Mansfield and Forsythe (1993)	Antibody coated magnetic particles
Biophotometry	Thomas *et al.* 1985)	Detection of microbial presence of optical density measurements
Radiometry	Stevens (1983)	Detection using radioisotope markers
Gas detection	Sawhney *et al.* (1986)	Detection of gases produced by microbial growth in media
Biosensors	Scott (1985)	A variety of detection principles, e.g. glucose oxidase
Filtration	Davis *et al.* (1984)	Microbial detection by concentration on filters

caused contamination problems in the past (see chapter 2). *Pseudomonas* is probably the most commonly isolated group of organisms (McCarthy 1980), particularly *Pseudomonas fluorescens* and *P. cepacia*, although in some instances difficulty in identification can make accurate categorization a problem. Many of these micro-organisms have very simple growth requirements and are adaptable to a wide range of environments (Stanier *et al.* 1966). As a result their detection by new methodologies must take into account the conditions in which they are found.

Table 14.3 Micro-organisms which cause contamination problems in pharmaceuticals, cosmetics and toiletries

Main groups	
Bacteria	*Pseudomonas*
	Enterobacter
	Klebsiella
	Bacillus spores
Less frequently isolated	
Bacteria	*Serratia*
	Citrobacter
	Acinetobacter
	Proteus
	Staphylococcus
Yeast	*Candida*
Fungi	*Aspergillus*
	Penicillium

14.2 Methods

14.2.1 *Electrical methods*

The growth of micro-organisms in media results in a change in the electrical conductivity of the media. These changes occur in the medium owing to metabolic activity of the micro-organisms. The detection of electrical changes, whether measured as impedance, conductance or capacitance, is the underlying principle of the electrical detection methods. In order to achieve this electrical change, a threshold concentration of micro-organisms is required and this has been found to be approximately 10^6 colony-forming units per ml (cfu ml^{-1}) (Cady *et al.* 1978).

This threshold concentration can be reached by incubating samples in media in special containers where the electrical response can be continuously monitored. The media used for sample enrichment and microbial detection will depend on the type or types of organisms under investigation. Media must be carefully chosen to optimize method sensitivity. A number of broths specifically designed for use with conductance/impedance methods are now available.

The time taken to reach the threshold concentration is known as the detection time and this will depend on the initial concentration of micro-organisms in the sample and their growth rate in the medium. The main instrument manufacturers are Bactomatic, Malthus and Don Whitley. Such systems represent highly automated computer-backed techniques capable of handling more than 500 samples simultaneously (Kahn and Firstenberg-Eden 1984, Anon. 1980, Levett 1986, Kell and Davey 1990).

All these systems have been evaluated for toiletry QA by the present authors. From a wide range of growth media selected for test, brain heart infusion broth (Oxoid), plate count agar and impedance broth (Don Whitley) were found to give good results with contaminated shampoo, toothpaste, cream and conditioner samples. Time-to-detection varied depending on the product and organism concerned (7–27 h).

The major difficulty with this technique is the detection of non-fermentative Gram-negative bacteria. These produce little change in the electrical characteristics of the media and therefore give very long detection times offering no saving over traditional techniques. In addition many cosmetics, toiletries and pharmaceuticals interfere with the electrical methods as the raw materials themselves are often already highly charged. Both these problems can be overcome by use of indirect methods. These methods involve the measurement of CO_2 evolved by the culture as the organisms grow. The CO_2 is absorbed into an alkaline solution whose change in conductance is monitored. This allows organisms to be grown in any liquid or solid media and in the presence of highly charged raw materials (Owens *et al.* 1989).

Impedance tests for specific organisms using selective impedance broths are used in the food industry (Smith *et al.* 1991) and have been shown to be highly reproducible and results correlate well with initial TVC. However, for general microbiological monitoring of toiletries and cosmetics the unknown genus of the contaminating micro-organisms makes correlation with actual TVCs of the products difficult. Connolly *et al.* (1993, 1994) have found success using impedance for preservative efficacy testing of these products. Good correlations were found with TVCs of untreated suspensions and impedance of specific organisms and between TVCs of preservative treated suspensions and impedance. However, the correlations were significantly different from one another and from species to species.

The range of micro-organisms that may be isolated from non-sterile pharmaceuticals or toiletries will demonstrate various growth rates as well as different product contamination levels. With two variables, therefore, any method involving a microbial growth step in the analysis may not correlate with the original level of contamination in the product. In this case the method can only be used as a qualitative test to detect presence or absence of viable contaminants.

14.2.2 *Direct epifluorescent filtration technique*

This technique depends on the observation that viable acridine orange-stained bacteria tend to fluoresce orange/red. Micro-organisms concentrated by filtration from the original sample are stained with acridine orange and viable cells counted under a fluorescent microscope. The method can detect greater than 10^3–10^4 micro-organisms ml^{-1} of product (Pettipher and Rodrigues 1981) although this sensitivity greatly depends on the quantity of product that can be filtered. The direct epifluorescent filtration technique (DEFT) has a number of distinct advantages over many other methods (Fung 1991). Firstly, it is a truly rapid technique, the method taking less than 1 h for sample analysis. Secondly, it is a quantitative technique (Rodrigues and Kroll 1985) although there may be problems in differentiating viable from non-viable cells in some samples (Rodrigues and Kroll 1986). The DEFT has been investigated for milk analysis (Pettipher *et al.* 1980), milk product analysis (Pettipher and Rodrigues 1981) and food analysis (Pettipher and Rodrigues 1982). The DEFT has also been evaluated for the detection of bacterial contaminants in intravenous fluids where detection down to 25 organisms ml^{-1} was obtained by increasing the sample volume filtered (Denyer and Ward 1983). Recently a method has been developed to overcome the problem of non-viable cells staining orange/red. This involved filtration of the sample followed by a short incubation to allow microcolonies to form. The sample was stained and the viable microcolonies were then counted (Newby 1991).

Most problems encountered with the DEFT relate to the filtration of samples. Whilst liquids of low viscosity can be easily filtered, other products and raw materials require pretreatment and/or prefiltration to get the sample through the filters. This makes the method impracticable for some routine operations where product viscosity or composition prevents filtration. Liquids of viscosity up to and including that of most shampoo formulations can be filtered either directly or with a predilution step, although it was noted with a shampoo formulation that the bacteria fluoresced green on the filter and not orange as expected of viable micro-organisms (Kroll, personal communication). This observation may well reflect the poor metabolic state of the micro-organisms in the formulation, i.e. they were under stress and nutrient depleted in the product. Filtration of large volume samples may also leave significant particulate debris on the filter surface which may hinder accurate counting (Connolly *et al.* 1993).

14.2.3 Microcalorimetry

Microbial catabolic activity produces heat that can be measured on a sensitive microcalorimeter (Suurkuusk and Wadsö 1982, Beezer 1980). Microbial heat production can be detected by flowing the sample continuously through the calorimeter or by suspending the sample in a growth medium inside a sealed metal ampoule in the calorimeter (Beezer *et al.* 1980, Harju-Jeanty 1982). Unfortunately, continuous flow is very dependent on the viscosity of the formulation and it was found that even shampoo formulations could not be continuously monitored unless they were prediluted. Investigations of the application of microcalorimetry to monitoring of toiletries are illustrated in Fig. 14.1. The microcalorimeter was operated in ampoule mode, formulations being diluted in tryptone soya broth (Oxoid) and compared against the heat evolved from the sterile broth. It was found that the machine could detect micro-organisms within 24 h. The detection time could not be correlated with the number of microbes in the formulation because the detection time was related to the growth rate of the particular isolates rather than the initial number of viable cells. For example, although the two shampoos in Fig. 14.1 had equivalent starting concentrations of viable cells, the contamination of 2.2×10^4 cfu ml^{-1} was detected after only 4 h, whereas the contamination of 1.8×10^4 cfu ml^{-1} took 11–12 h for detection. Analysis of these results indicated that the number of micro-organisms required to give a thermal response, i.e. the threshold concentration, was between 10^4 cfu ml^{-1} and 10^5 cfu ml^{-1}. This means that products monitored continuously in flow microcalorimetry would need to be quite heavily contaminated before a thermal response was obtained.

Another testing feature of microcalorimetry is that the thermal traces produced are peculiar to each micro-organism. The traces produced in our experiments (Fig. 14.1) were reproducible, provided that constant conditions were maintained, and these curves may offer information about the type of micro-organism and possibly even an identification. Insufficient work has carried out to resolve this issue conclusively although this concept has been reported previously (Monk and Wadsö 1975).

14.2.4 Flow cytometry

Methods for counting and sizing of micro-organisms with electronic transducers have been available for some time (Kubitschek 1969). Flow cytometry is both rapid and

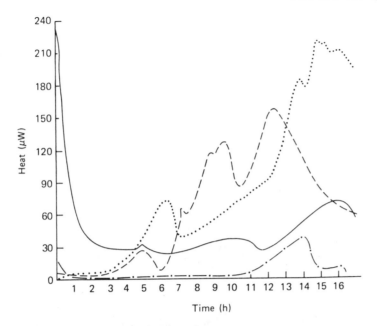

Figure 14.1 Microcalorimetry curves produced on a BioActivity Monitor (LKB) of four toiletry samples diluted 1:10 in tryptone soya broth for various initial micro-organism concentrations and various samples: ——, 1.1×10^5 cfu ml^{-1}, conditioner; ······, 7.7×10^5 cfu ml^{-1}, conditioner; ---, 2.2×10^4 cfu ml^{-1}, shampoo; —·—; 1.8×10^4 cfu ml^{-1}, shampoo

quantitative although, being a total count, dead micro-organisms are also detected. Micro-organisms are detected by passing the sample, diluted in a conducting electrolyte fluid such as saline, through a small aperture across which an electric current flows. Changes in electrical resistance in the aperture relate to the passage of particles, e.g. micro-organisms. One of the major disadvantages of this method is that the aperture must be very small for microbial detection and may be easily blocked by microbial aggregates. The Orbec system with its wider aperture has been evaluated with a few products to investigate its applicability in our laboratory. Figs 14.2 and 143 show the response that can be achieved with a shampoo with and without contaminating bacteria. A clear indication can be obtained of the presence of micro-organisms. Recent advances have to some extent overcome the problem of differentiating viable from non-viable cells. Kaprelyants and Kell (1992) have reported the use of a dye, Rhodomine 123, to indicate not only viable and non-viable cells in flow cytometer studies but also those cells which are capable of being resuscitated.

Chemunex (Cambridge UK) have developed a similar system involving a substrate which can be metabolized and then detected by the flow cytometer. The method can reportedly detect organisms in a product which contains significant particulate matter, a problem which has affected other systems.

14.2.5 *Biophotometry*

Detection of microbial growth by either optical density or light transmission is a well-established technique (Cruickshank 1965). Instrumentation, e.g. Abbott MS2 and

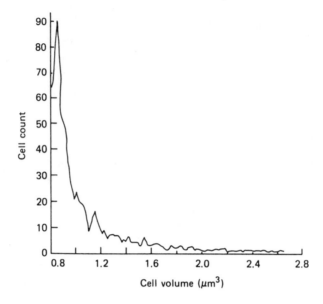

Figure 14.2 Uncontaminated shampoo diluted 0.1 ml in 20 ml of electrolyte and analysed in the Orbec counter

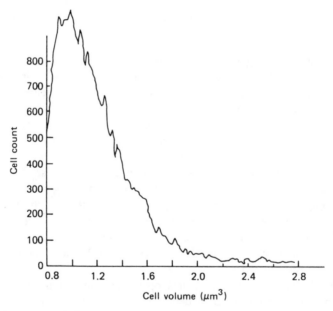

Figure 14.3 Contaminated shampoo (Gram-negative pseudomonad) diluted 0.1 ml in 20 ml of electrolyte and analysed in the Orbec counter

Bioscreen, has been designed for monitoring biophotometric changes and detection data indicated that detection times are similar to those obtained with the electrical methods, i.e. the detection thresholds are similar. The major problem with this technique is the opacity of some formulations and raw materials which reduces the sensitivity of the method or masks detection altogether. Again, because this method relies on a growth phase for detection, it means that the methods can only be qualitative.

14.2.6   *Limulus amoebocyte lysate (LAL)*

The *Limulus* amoebocyte lysate test detects lipopolysaccharide, a product that is produced by all Gram-negative bacteria (Bergan 1981). Lipopolysaccharide will gel *Limulus* amoebocyte lysate, which is derived from the amoebocytes of the horseshoe crab, *Limulus polyphemus*, and the gel can be detected either biophotometrically (Jorgensen and Alexander 1981), colorimetrically (Ogawa and Kanoh 1984) or chromogenically (Friberger 1987). A relationship has been shown between this test and pyrogen testing (Kanoh and Kawasaki 1980) and the only disadvantages of the technique are firstly that the sensitivity is approximately 10^5 cfu ml^{-1} and secondly that alternative tests are required to detect other bacteria and moulds.

The LAL test is now used extensively for the detection of bacterial endotoxins in the manufacture of sterile pharmaceutical products, when, not only screening raw materials and monitoring production processes and high-purity water systems, but also when releasing selected sterile products.

14.2.7   *Adenosine triphosphate detection*

This method involves the detection of adenosine triphosphate (ATP) in micro-organisms. ATP is the main chemical store of energy in all types of cells and so the method should detect all types of micro-organisms, as mentioned in section 14.1.3.

In other industrial areas, e.g. the food industry, the presence of somatic ATP makes the detection of micro-organisms more difficult (Stannard 1984) but this is not generally a problem with cosmetics and toiletries.

ATP detection has been investigated for sterility testing of pharmaceuticals (Bussey and Tsuji 1986) and for bacteriological screening of toiletries and cosmetics (Borovian 1991, Nielsen and Van Dellen 1989) and as such may have application across all three industries. Some general reviews of the technique have been reported by Stanley (1993), Stanley *et al.* (1989) and Szalay *et al.* (1993). The method for ATP analysis involves the firefly bioluminescence reaction (Fig. 14.4). This reaction uses one molecule of ATP to oxidize the substrate (firefly luciferin) in the presence of an enzyme (firefly luciferase) to produce one photon of light. There is a range of ATP instrumentation available that will monitor the light output from the reaction. Jago *et al.* (1989) reviewed the performance of ten luminometers. There have been a number of advances in recent years particularly with respect to automation and computer control and also with respect to portable instruments.

The general method for ATP analysis involves extraction of ATP from micro-organisms in the sample with a release agent. The ATP can then be reacted with luciferin and luciferase and the light output monitored. In applying this method to different types of material, pretreatment of samples may be required to achieve maximum sensitivity. In applying the method to cosmetics and toiletries there are a number of limitations which are:

Micro-organisms
$$\downarrow$$

$$\text{ATP+luciferin+luciferase} \xrightarrow{\text{Mg}^{2+}} \text{luciferase}-\text{luciferin}-\text{AMP}+\text{PP}$$

$$\text{luciferase}-\text{luciferin}-\text{AMP}+\text{O}_2 \rightarrow \text{luciferase}+\text{AMP}+\text{CO}_2+\text{oxyluciferin}+\text{hv}$$
$$\downarrow$$
$$\text{to light detector}$$

Figure 14.4   The ATP bioluminescence reaction (McElroy and DeLuca 1981): AMP, adenosine monophosphate; PP, pyrophosphate

(1)   The ATP reaction is easily quenched by a wide variety of chemicals. Table 14.4 illustrates the effect of a number of diluents, the quenching being quite significant even with simple chemicals such as saline. Since quenching occurs with many formulation ingredients, this effect must be overcome by dilution of samples before performing the ATP analysis or by centrifugation of the incubated sample. The micro-organisms can be resuspended in sterile distilled water or a buffer giving low quenching (Nielsen and Van Dellen 1989).

(2)   Under ideal conditions with good quality reagents and instruments, it is possible to detect 10^2 cfu g^{-1}; however in practice during routine analysis in the presence of broths, diluted raw materials etc., the method is only sensitive to approximately 10^4 cfu g^{-1} (Manger 1984). As such, this sensitivity is too low for the analysis of most cosmetics and toiletries where the EEC recommended limit is not greater than 10^3 micro-organisms g^{-1} of product (Spooner 1977). In this situation, pretreatment of samples is required to increase the number of organisms per gram or millilitre of product.

(3)   As mentioned in section 14.2.2, the metabolic activity of contaminants in toiletries is often quite low. An enrichment step will increase microbial metabolic activity of the cells, thereby elevating the ATP levels.

Experiments conducted by the authors have shown that:

(1)   Pre-enrichment media must be carefully controlled to have a low background ATP level.

(2)   Sample dilution in the pre-enrichment medium and medium dilution prior to ATP assay must be optimized for each product type to maximize detection sensitivity.

(3)   Shaking pre-enrichments (200 rev min^{-1}) during incubation increased luminometer readings 30-fold.

(4)   Pre-enrichment for a minimum of 18 hours was required to detect contamination levels of 10 cfu g^{-1} in product samples.

(5)   Detection reagents can be obtained commercially with a variety of sensitivity levels and these can be further manipulated by changing the levels of chemical constituents in the mixture.

Table 14.4   Quenching of bioluminescence by media and diluents

Medium or diluent	Unquenched reaction[a] (%)
Ringers saline	11
Nutrient broth No. 1	8
Brain heart infusion	3

[a] Unquenched reaction obtained using sterile deionized water as the diluent.

An example of a detection methodology for ATP is summarized in Fig. 14.5. This methodology was found to be capable of reproducibly detecting 5×10^{-12} M ATP. On the assumption of a bacterial ATP content of approximately 1×10^{-15} g cell^{-1}, the method had a lower detection sensitivity of 2.5×10^3 bacteria ml^{-1}. For yeast, on the assumption of an ATP content of approximately 2×10^{-14} g cell^{-1}, the sensitivity was 1.3×10^2 yeast ml^{-1}.

As it stands, the method was capable of detecting 10 cfu g^{-1} or more in 18 h or inversely of clearing satisfactory product in the same time. It was not a quantitative method (i.e. it did not give an estimate of the initial product contamination) since as with electrical methods, the time for ATP detection depends not only on the initial amount of contamination but also on the lag phase for recovery of injured cells and on the microbial growth rate in the enrichment stage. Despite the automation, the method required some microbiological skill for the enrichment stage. It has been noted on a number of occasions that ATP analysis proved to be a better indicator of product contamination than the traditional plating techniques which failed to produce colonies within the specified incubation period.

14.2.8   Other methods

Radiometry has always been a problem as a rapid detection technique for use in industry because it involves the use of radioisotopes. It has been suggested as a possible rapid method for sterility testing, microbial assays and microbial limit tests for pharmaceuticals (Anderson 1986). The method is based on the detection of ^{14}C-labelled carbon dioxide produced from a ^{14}C substrate by metabolizing micro-organisms. Recently, this method has been superseded by an infrared carbon dioxide detector technique which may have a wider acceptability. The method still has a delay whilst microbial growth occurs and so, like many other methods, it is faster than traditional methods rather than rapid, and probably qualitative rather than quantitative.

The detection of microbial enzymes has rapid method possibilities provided that all the micro-organisms being investigated produce the appropriate enzyme. Cytochrome *c* oxidase (Kroll 1985) and catalase (Charbonneau *et al.* 1975) are two enzymes that can be used for microbial detection although, as with many of the other techniques, sensitivity may be the largest problem with exploiting this approach unless a microbial enrichment step is used. A recent patent has described the use of DMSO reduction to detect the presence of micro-organisms in cosmetics and pharmaceutical samples. The dimethyl sulphide produced by the action of the organisms can be measured in the gas phase by gas chromatography after incubation for 2–3 hours (Kleiner 1993).

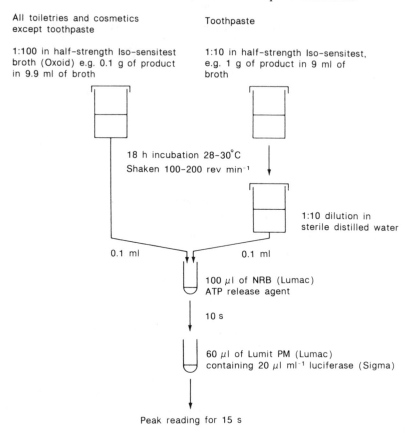

All toiletries and cosmetics except toothpaste

1:100 in half-strength Iso-sensitest broth (Oxoid) e.g. 0.1 g of product in 9.9 ml of broth

Toothpaste

1:10 in half-strength Iso-sensitest, e.g. 1 g of product in 9 ml of broth

18 h incubation 28-30°C
Shaken 100-200 rev min⁻¹

1:10 dilution in sterile distilled water

0.1 ml

0.1 ml

100 μl of NRB (Lumac)
ATP release agent

10 s

60 μl of Lumit PM (Lumac)
containing 20 μl ml⁻¹ luciferase (Sigma)

Peak reading for 15 s

Figure 14.5   An ATP method summary

Deoxyribonucleic acid probes and enzyme-linked immunosorbent assay techniques are more applicable to rapid detection and identification of specific micro-organisms which may have most relevance to the pharmaceutical industry (Grant *et al.* 1993, Lampel *et al.* 1990). However, there have been some rapid advances in this field which may be applicable in the cosmetics, toiletries as well as pharmaceutical industries in the future. Brooks and Kroll (1992) have developed DNA and RNA probes of Listeria and Carnobacteria. These probes are linked to enzymes whose products can be detected electrically hence giving direct signals from unenriched food samples. It is reported that 10^2 cfu g^{-1} can be detected in less than 2 hours. Similarly Brooks *et al.* (1992) have reported enzyme-linked antibodies for salmonella detection which can be directly monitored by electrical signals. A count of 10^4 cfu ml^{-1} was detected in 4 hours in food samples, though 5 cfu ml^{-1} were detected following a 6-h enrichment step.

These technologies have been further extended by use of magnetic beads bound to antibodies and RNA probes. The organisms became bound to the complex and could then be separated from the remainder of the media, thus giving a concentration step (Mansfield and Forsythe 1993).

Whilst numerous other methods have been postulated as rapid alternatives to traditional techniques, e.g. biosensors (Gronow 1984), potentiometric detection (Junter 1984) and electron transfer coupling (Kroll 1986), they are at relatively early development stages and their applicability in the future to the pharmaceutical, cosmetic and toiletries industries is difficult to assess.

14.4   Identification systems

This section describes the newer microbial identification techniques which are not routinely used in the pharmaceutical, cosmetic or toiletry industries at the present time. Some of the methods introduced earlier in this chapter allow not only microbial detection but also identification, e.g. deoxyribonuclei acid probes. The miniaturized identification kits, e.g. API and Enterotube, are now routinely used in many laboratories and these are being supplemented and updated by automated methods based on assimilation tests (Freney *et al.* 1984, Palmieri *et al.* 1988) (e.g. ATB system (API)), antibiotic susceptibility methods (Waterworth 1983, Gavan and Town 1970) (e.g. Microscan system (American Micro-Scan)) or biochemical tests (Kelly and Latimer 1980) (e.g. Auto-Microbic system (Vitek systems)), (Knight *et al.* (1993) (e.g. Biolog Identification System).

Other automated methods are based on a number of different detection principles and can be summarized as follows.

14.4.1   *Electrophoresis*

Microbial identification is obtained by electrophoresing radiolabelled proteins obtained by growing the micro-organisms in ^{35}S methionine. This produces a banded pattern of proteins that is unique to each species of micro-organism (Anon. 1985).

14.4.2   *Gas chromatography*

Microbial fatty acids can be separated and identified using gas chromatography. The resulting 'fingerprint' is characteristic of each microbial species and can be used to obtain a computer identification (Miller and Berger 1985).

14.4.3   *Pyrolysis mass spectrometry*

Microbial isolates are heated in a vacuum, which causes thermal breakdown of the cells. The resultant gas phase is analysed directly in a mass spectrometer and the spectra obtained can give a microbial identification when compared to spectra of known micro-organisms (Anon. 1982, Ottley and Maddock 1986).

Although all these methods can represent faster identification methods than are currently used, they also represent a high capital investment in the hardware. It is also often difficult to develop a comprehensive database of microbial identities for these systems because of the non-standard nature of many microbial isolates from pharmaceuticals, cosmetics and toiletries.

14.5 Conclusions

Current methodology is often too slow to act as an effective means of MQA. The rapid methods now on the market are capable of reducing analysis time although the term 'faster analysis' would be more appropriate than 'rapid' for most techniques. Additionally, although most new methods can replace traditional techniques, they must go through a development stage to adapt them to the requirements and standards of each industry. This development phase may be too long and expensive for the involvement of some companies. Furthermore, having developed and adapted a method, problems may arise in establishing its use with regulatory authorities. The difficulties that this entails has reduced methods implementation in the pharmaceutical industry. The main criteria on which to base a change to a new method can be summarized as follows.

(1) How quickly are results required? If results are required immediately, then many new methods are unsuitable because they require an incubation step.

(2) What quality of data is required? Many of the methods will fulfil a qualitative role but few are truly quantitative.

(3) What level of automation is required? To a certain extent, this component is also linked to (4): the greater the automation, the higher is the price. The level of automation will largely be dictated by the sample throughput required and the number of staff available.

(4) What is the cost? Again this is a factor controlled by the sample throughput but also by the degree of automation required.

Having selected a method, it is only a matter of time to develop the technique for routine operations. Additional work may be required in certain applications in the pharmaceutical industry to obtain acceptance of the method by the regulatory authorities.

References

ANDERSON, B. (1986) Pharmaceutical applications for the Bactec 460. *J. Appl. Bacteriol.* **61** v–vi.
ANON. (1980) Machine speeds up microbe tests. *New Scientist* December 711.
    (1982) Chocolate firm choose microbe detector for a new centre. *New Scientist* February 504.
    (1985) Scanning for faster microbiology. *Lab. Pract.* January 29–30.
BAIRD, R. M. (1990) Monitoring microbiological quality: conventional testing methods. In: Denyer, S., and Baird, R. M. (eds). *Guide to microbiological control in pharmaceuticals*. Ellis Horwood, Chichester.
BEEZER, A. E. (1980) *Biological microcalorimetry*. Academic Press, London.
BEEZER, A. E., MILES, R. J., SHAW, E. J., and VICKESTAFF, L. (1980) Antibiotic bioassay by flow microcalorimetry. *Sep. Exp.* **36** 1051–1052.
BERGEN, T. (1981) Rapid methods for detection of aerobic bacteria in blood and spinal fluid. In: Tilton, R. C. (ed.), *Rapid methods and automation in microbiology*. American Society for Microbiology, Washington, DC, pp. 36–40.
BOROVIAN, G. E. (1991) Bioluminescence method for the rapid detection of molds in the quality control of cosmetics, pharmaceutical products and raw materials. *Proc. 91st Gen. Meeting Am. Soc. Microbiol.*, Dallas, May 1991.
BROOKS, J. L., and KROLL, R. G. (1992) Electrical detection of enzyme labelled oligonucleotide

probes to Listeria and Carnobacterium in pure cultures and in foods. *J. Microbiol. Methods* **16** 125–137.

BROOKS, J. L., MIRHABIBOLLAHI, B., and KROLL, R. G. (1992) Experimental enzyme-linked amperometric immunosensors for the detection of salmonellas in foods. *J. Applied. Bact.* **73** 189–196.

BUSSEY, D. M., and TSUJI, K. (1986) Bioluminescence for USP sterility testing of pharmaceutical suspension products. *Appl. Environ. Microbiol.* **51** 349–355.

CADY, P. DUFOUR, S. W., SHAW, J., and KRAEGAR, S. J. (1978) Electrical impedance measurements: rapid method for detecting and monitoring micro-organisms. *J. Clin. Microbiol.* **7** 265–272.

CHAPPELLE, E. W., and LEVING, V. (1968) Use of the firefly bioluminescent reaction for rapid detection and counting of bacteria. *Biochem. Med.* **2** 41–51.

CHARBONNEAU, R., THERRIEN, J., and GAGNON, M. (1975) Detection and measurement of bacterial catalase by the disc flotation method using the catalase meter. *Can. J. Microbiol.* **21** 580–582.

COHEN, E. (1979) In: *Biochemical applications of the horseshoe crab (Limulidae).* A. R. Liss, New York.

CONNOLLY, P., BLOOMFIELD, S. F., and DENYER, S. P. (1993) A study of the use of rapid methods for preservative efficacy testing of pharmaceuticals and cosmetics. *J. Applied Bact.* **75** 456–462.

—— (1994) The use of impedance for preservative efficacy testing of pharmaceutical and cosmetic products. *J. Appl. Bact.* **76** 68–75.

CRUICKSHANK, R. (1965) *Medical microbiology*, 11th edn. E. and S. Livingstone, Edinburgh, pp. 872–873.

DAVIS, J. R., STAGER, C. E., and ARAJ, G. F. (1984) Clinical laboratory evaluation of a bacteriuria detection device for urine screening. *Am. J. Clin. Pathol.* **81** 48–53.

DENYER, S. P., and WARD, K. H. (1983) A rapid method for the detection of bacterial contaminants in intravenous fluids using membrane filtration and epifluorescent microscopy. *J. Parenter Sci. Technol.* **37** 156–158.

FITTER, S., HEUZENROEDER, M., and THOMAS, C. J. (1992) A combined PCR and selective enrichment method for rapid detection of Listeria monocytogenes. *J. Applied Bact.* **73** 53–59.

FORREST, W. W. (1972) Microcalorimetry. In: Norris, J. R., and Ribbons, D. W. (eds), *Methods in microbiology*, Vol. 6b. Academic Press, London, pp. 285–318.

FRENEY, J., LABAN, P., DESMONCEAUX, M., GAYRAL, J. P., and FLEURETTE, J. (1984) Differentiation of Gram-negative rods other than Enterobacteriaceae and Vibrionaceae by a micromethod for determination of carbon substrate assimilation. *Zentralbl. Bakteriol. Hyg. A* **258** 198–212.

FRIBERGER, P. (1987) A new method of endotoxin determination. *ICPR* July/August 34–41.

FUNG, D. Y. C. (1991) Rapid methods in applied microbiology: an update. In *Proc. Biodeterior. Biodegrad. Symp.* **8** 19–34.

GAVAN, T. L., and TOWN, M. A. (1970) A microdilution method for antibiotic susceptibility testing. *Am. J. Clin. Pathol.* **53** 880.

GONZALEZ, I., MARTIN, R., GARCIA, T., MORALES, P., SANZ, B., and HERNANDEZ, P. E. (1993) A sandwich enzyme-linked immunosorbent assay (ELISA) for detection of Pseudomonas fluorescens and related psychotrophic bacteria in refrigerated milk. *J. Applied Bact.* **74** 393–401.

GRANSTROEM, M., WRETLIND, B., MARKMAN, O. R. P., and VASIL, M. L. (1984) Enzyme linked immunosorbent assay for detection of antibodies to *Pseudomonas aeruginosa* exproteins. *Eur. J. Clin. Microbiol.* **4** 197–200.

GRANT, K. A., DICKINSON, J. H., PAYNE, M. J., CAMPBELL, S., COLLINS, M. D., and KROLL, R. G. (1993) Use of the polymerase chain reaction and 16S rRNA sequences for the rapid detection of *Brochothrix* spp. in foods. *J. Applied Bact.* **74** 260–267.

231

GRIFFITHS, M. W., and PHILLIPS, J. D. (1982) Identification of bacteria of dairy origin using miniaturised test systems. *J. Appl. Bacteriol.* **53** 343–350.

GRONOW, M. (1984) Biosensors. *Trends Biochem. Sci.* **9** No. 8, 336–340.

HARJU-JEANTY, P. (1982) Microcalorimetry, a new technique for studying slime problems in papermaking waters. *Appita,* **36** 26–31.

HOLMES, B., WILCOX, W. R., LAPAGE, S. P., and MALWICK, H. (1977) Test reproducibility of API 20E, Enterotube and Pathotec systems. *J. Clin. Pathol.* **30** 381–387.

HOOPER, W. L. (1981) Microbiological testing of non-sterile pharmaceuticals. *J. Appl. Bacteriol.* **51** vi–vii.

JAGO, P. H., SIMPSON, W. J., DENYER, S. P., EVANS, A. W., GRIFFITHS, M. W., HAMMOND, J. R. M., INGRAM, T. P., LACEY, R. F., MACEY, N. W., McCARTHY, B. J., SALISBURY, T. T., SENIOR, P. S., SIDOROWICZ, S., SMITHER, R., STANFIELD, G., and STANLEY, P. E. (1989) An evaluation of the performance of ten commercial luminometers. *J. Biolumin. Chemiluminescence,* **3** 131–145.

JORGENSEN, J. H., and ALEXANDER, G. A. (1981) Automation of the *Limulus* amoebocyte lysate test by using the Abbott MS-2 microbiology system. *Appl. Environ. Microbiol.* **41** 1316–1320.

JUNTER, G. A. (1984) Potentiometric detection and study of bacterial activity. *Trends Anal. Chem.* **3** No. 10, 253–259.

KAHN, P., and FIRSTENBERG-EDEN, R. (1984) A new cosmetic sterility test. *Soap, Cosmet., Chem. Spec.* **60** No. 5, 46–48, 101.

KANOH, S., and KAWASAKI, H. (1980) Studies on the relationship between the pyrogen test and *Limulus* test. *Bull. Natl. Inst. Hyg. Sci. Tokyo* **55** No. 98, 76–80.

KAPRELYANTS, A. S., and KELL, D. B. (1992) Rapid assessment of bacterial viability and vitality by rhodamine 123 and flow cytometry. *J. Applied Bact.* **72** 410–422.

KELL, D. B., and DAVEY, C. L. (1990) Conductimetric and impedimetric devices. In; *Biosensors – a practical approach.* Oxford University Press, Oxford, pp. 125–154.

KELLY, M. T., and LATIMER, J. M. (1980) Comparison of the automicrobic system with API, Enterotube, Micro-ID, Micro-Media systems and conventional methods for identification of Enterobacteriaceae. *J. Clin. Microbiol.* **12** 659–662.

KLEINER, D. (1993) Rapid determination of microbiological activity – by adding DMSO to the foodstuff, cosmetic or pharmaceutical sample and analysing the dimethyl sulphide produced in the gas phase. Patent; DE4136390 (German).

KNIGHT, G. C., McDONNELL, S. A. SEVIOUR, R. J., and SODDELL, J. A. (1993) Identification of Acinetobacter isolates using the Biolog identification system. *Lett. Applied Microbiol.* **16** 261–264.

KROLL, R. G. (1985) The cytochrome *c* oxidase test for the rapid detection of psychrotrophic bacteria in milk. *J. Appl. Bacteriol.* **59** 137–141.

   (1986) The detection of cytochrome *c* oxidase and electron transfer. *J. Appl. Bacteriol.* **61** No. 6, iv.

KROLL, R. G., and RODRIGUES, U. M. (1986) Prediction of the keeping quality of pasteurised milk by the detection of cytochrome *c* oxidase. *J. Appl. Bacteriol.* **60** 21–27.

KROLL, R. G., PATCHETT, R. A., LEAROYD, E., and THURSTON, C. F. (1989) Detection of electron transfer for the assessment of bacterial contamination. In: Stannard, C. J., Petitt, S. B., and Skinner, F. A. (eds), *Rapid microbiological methods for foods, beverages and pharmaceuticals.* Soc. Applied Bact. Tech. Series No. 25. Blackwell Scientific, London, pp. 227–240.

KUBISCHEK, H. E. (1969) Counting and sizing micro-organisms with the Coulter counter. In: Norris, J. R., and Ribbons, D. W. (eds), *Method in microbiology*, Vol. 1. Academic Press, London, pp. 593–610.

LAMPEL, J. A., JAGOW, J. A., TRUCKSESS, M., and HILL, W. E. (1990) Polymerase chain reaction for detection of invasive *Shigella flexneri* in food. *Applied Environ. Microbiol.* **56** 1536–1540.

LEVETT, B. C. J. (1986) Impedance microbiology – its basic principles and application to industrial testing. *Jpn. J. Food Microbiol.* **3** No. 1, 69–73.

MADDEN, J. M. (1984) Microbiological methods for cosmetics. In: Kabara, J. J. (ed.), *Cosmetic and drug preservation*. Marcel Dekker, New York, pp. 573–603.

MCCARTHY, T. J. (1980) Microbiological control of cosmetic products. *Cosmet. Toilet.* **94** 23–27.

MCELROY, W. D., and DELUCA, M. (1981) The chemistry and applications of firefly luminescence. In: DeLuca, M. A., and McElroy, W. D. (eds), *Bioluminescence and chemiluminescence*. Academic Press, New York, pp. 179–186.

MANGER, P. A. (1984) Preliminary evaluation of the Lumac Biocounter M2010 system, measuring bacterial ATP in food, using a bioluminescent technique. In: Kricka, L. J., Stanley, P. E., Thorpe, G. H. G., and Whitehead, T. P. (eds), *Analytical applications of bioluminescence and chemiluminescence*. Academic Press, New York, pp. 57–60.

MANSFIELD, L. P., and FORSYTHE, S. J. (1993) Immunomagnetic separation as an alternative to enrichment broths for Salmonella detection. *Lett. Applied Bact.* **16** 122–125.

MILLER, L., and BERGER, T. (1985) Bacteria identification by gas chromatography of whole cell fatty acids, Hewlett-Packard Application Note. Hewlett-Packard, pp. 228–241.

MONK, P., and WADSÖ, I. (1975) The use of microcalorimetry for bacterial classification. *J. Appl. Bacteriol.* **38** 71–74.

NEWBY, P. (1991) Analysis of high quality pharmaceutical grade water by a direct epifluorescent filter technique – microcolony method. *Lett. Applied Microbiol.* **13** 291–294.

NIELSEN, P., and VAN DELLEN, E. (1989) Rapid bacteriological screening of cosmetic raw materials by using bioluminescence. *J. Assoc. Anal. Chem.* **72** No. 5, 708–711.

NORD, C. E., LINDBERG, A. A., and DAHLBACK, A. (1974) Evaluation of five kits, API, Autotab, Enterotube, Pathotec and R/B for identification of Enterobacteriaceae. *Med. Micro. Immun. Serol.* **159** 211–220.

OGAWA, Y., and KANOH, S. (1984) Enhancement of endotoxicity and reactivity with carbocyanine dye by sonication of lipopolysaccharide. *Microb. Immunol.* **28**, 1313–1323.

OTTLEY, T. W., and MADDOCK, C. J. (1986) The use of pyrolysis mass spectrometry. *Lab. Pract.* October 53–55.

OWENS, J. D., THOMAS, D. S., THOMPSON, P. S., and TIMMERMAN, J. W. (1989) Indirect conductimetry: a novel approach to the conductimetric enumeration of microbial population. *Lett. Applied Microbiol.* **9** 245–249.

PALMIERI, M. J., CARITO, S. L., and MEYER, R. F. (1988) Comparison of rapid NFT and API 20E with conventional methods for identification of Gram-negative nonfermentative Bacilli from pharmaceuticals and cosmetics. *Applied Environ. Microbiol.* **54** 2838–2841.

PETTIPHER, G. L. (1983) *The direct epifluorescent filtration technique*. Research Studies Press, Letchworth.

PETTIPHER, G. L., and RODRIGUES, U. M. (1981) Rapid enumeration of bacteria in heat-treated milk and milk products using a membrane filtration epifluorescence microscopy technique. *J. Appl. Bacteriol.* **50** 157–166.

(1982) Rapid enumeration of micro-organisms in foods by the direct epifluorescent filter technique. *Appl. Environ. Microbiol.* **44** 809–813.

PETTIPHER, G. L., MANSELL, R., MCKINNON, C. H., and COUSINS, C. M. (1980) Rapid membrane filtration-epifluorescent microscopy technique for direct enumeration of bacteria in raw milk. *Appl. Environ. Microbiol.* **39** 423–429.

RODRIGUES, U. M., and KROLL, R. G. (1985) The direct epifluorescent filter technique (DEFT): increased selectivity, sensitivity and rapidity. *J. Appl. Bacteriol.* **59** 493–499.

(1986) Use of the direct epifluorescent filter technique for the enumeration of yeasts. *J. Appl. Bacteriol.* **61** 139–144.

SAWHNEY, D., HINDER, S. M., SWAINE, D., and BRIDSON, E. Y. (1986) Novel methods for detecting micro-organisms in blood cultures. *J. Clin. Pathol.* **39** 1259–1263.

SCOTT, A. O. (1985) Biosensors: a tool for the food and beverage industries? *Lab. Pract.* June 39–43.

SMITH, P. J., BOLTON, F. J., and GAYNER, V. E. (1991) Reproducibility of electrical conductance assay results for the detection of salmonella. *Lett. Applied Microbiol.* **12** 78–80.

SPOONER, D. F. (1977) Microbiological aspects of the EEC Directive. *Cosmet. Toilet.* **92** 42–51.

STANIER, R. Y., PALLERONI, N. J., and DOUDOROFF, M. (1966) The aerobic pseudomonads: a taxonomic study. *J. Gen. Microbiol.* **43** 159–271.

STANLEY, P. E. (1993) Some views of the current status of ATP rapid methodology for use in food and beverage industries. In: Szalay, A. A., Kricka, L. J., and Stanley, P. E. (eds), *Bioluminescence and chemiluminescence, Proc. VII int. sym. bioluminescence and chemiluminescence*. John Wiley, Chichester, pp. 434–438.

STANLEY, P. E., McCARTHY, B. J., and SMITHER, R. (eds) (1989) *ATP luminescence: rapid methods in microbiology*. Soc. Applied Bacterial Tech. Series No. 26.

STANNARD, C. J. (1984) ATP assay as a rapid method for estimating microbial growth in foods. In: Kricka, L. J., Stanley, P. E., Thorpe, G. H. G., and Whitehead, T. P. (eds), *Analytical applications of bioluminescence and chemiluminescence*. Academic Press, London, pp. 53–56.

STEVENS, M. (1983) Rapid detection methods in microbiology. *Lab. Pract.* January 13–14.

SUURKUUSK, J., and WADSÖ, I. (1982) A multichannel microcalorimetry system. *Chem. Scr.* **20** 155–163.

SZALAY, A., KRICKA, L., and STANLEY, P. (eds) (1993) *Bioluminescence and chemiluminescence, Proc. VII int. symp. bioluminescence and chemiluminescence*, John Wiley, Chichester.

THOMAS, D. S., HENSCHKE, P. A., GARLAND, B., and TUCKNOTT, O. G. (1985) A microprocessor-controlled photometer for monitoring microbial growth in multi-walled plates. *J. Appl. Bacteriol.* **59** 337–346.

TOMKINS, L. S., MICKELSEN, P. A., and TROUP, N. (1985) DNA technology applied to the detection and epidemiology of enteric pathogens. In: Habermehl, K. O. (ed.), *Rapid methods and automation in microbiology and immunology*. Springer, Berlin, pp. 68–72.

WATERWORTH, P. M. 91983) Changes in sensitivity testing. *J. Antimicrob. Chemother.* **11** 1.

15

Microbiological contamination of manufactured products: official and unofficial limits

ROSAMUND M. BAIRD

Formerly of North East Thames Regional Health Authority Pharmaceutical Microbiology Laboratory, St. Bartholomew's Hospital, London EC1A 7BE[†]

Summary

(1)   The need to control microbiological contamination in pharmaceutical, cosmetic and toiletry products is well recognized.

(2)   Microbiological specifications have traditionally been based on one of two requirements: an acceptable total viable count or the absence of specified organisms in named products.

(3)   In the case of pharmaceuticals, microbiological control has been achieved through recommended pharmacopoeial standards whereas, with cosmetic and toiletries, it has been achieved through the adoption of guidelines agreed with industry.

(4)   The microbiological specifications should reflect the state of current knowledge and require revision from time to time.

15.1   Introduction

Although the importance of controlling microbiological contamination in pharmaceuticals, cosmetics and toiletries has been recognized for many years, the acceptance and implementation of microbiological standards for such products have, by comparison, developed more slowly. This chapter describes the way in which both official and unofficial microbiological standards have developed over the years, the detailed content of those standards, and how they have been implemented in practice.

15.2   Historical development of standards and guidelines

The first microbiological standards were based upon the criterion of sterility and initially applied only to that most critical group of pharmaceutical products, the injectable

[†] Present address: Summerlands House, Summerlands, Yeovil, Somerset BA21 3AL.

preparations. The sixth edition of the *British Pharmacopoeia (BP)* (Anon. 1932) intro-duced the requirement that injectable products were to be sterilized by various means (heating in an autoclave, tyndallization or filtration) and additionally were to pass a test for sterility. Surprisingly, the requirement for sterility was not applied to ophthalmic drops and lotions of the *British Pharmaceutical Codex* until a supplement was published in 1966 (Anon. 1966). Considerable documentary evidence had meanwhile accumulated from the early 1940s onwards, warning of the hazards of using contaminated eye drops (Allen 1959, Dale *et al.* 1959, Crompton 1962). In particular, fluorescein eye-drops had frequently been found to be a source of *Pseudomonas aeruginosa*, a notorious pathogen in ocular infections; indeed, some samples had been held responsible for the development of eye infections and subsequent loss of sight. In comparison, eye ointments were pre-sumed to be less susceptible to contamination owing to the nature of their composition, until an antibiotic eye ointment was reported in 1964 to be the source of an outbreak of *P. aeruginosa* eye infections in Sweden (Kallings *et al.* 1966). A requirement for sterility was subsequently applied to eye ointments listed in the *BP* of 1968 (Anon. 1968a).

Interestingly, despite the obvious similarities between ophthalmic solutions and con-tact lens solutions (essentially simple solutions instilled into eyes containing foreign bodies), the microbiological quality of the latter remained uncontrolled for a further decade. During this time, considerable evidence accumulated to suggest that the manufac-ture, preservation and performance of contact lens solutions were far from adequately controlled (Richards 1972, Norton *et al.* 1974, Davies and Norton 1975, Richardson *et al.* 1977). Finally, in 1980, contact lens solutions became a licensable product under the requirements of the Medicines Act (Anon. 1968b) and hence were required to be sterile. In contrast to pharmaceutical products, cosmetics used in and around the eye are not required to be sterile. Manufacturers do recognize the need, however, to formulate these cosmetics as well-preserved products and to manufacture them under conditions which will result in no more than a minimal microbial population. During subsequent storage and use this population should remain static or decline in number. Doubtless, as more information in time becomes available, the requirement for sterility may well be applied to additional groups of pharmaceutical preparations besides those mentioned above.

Unlike the requirement for sterility which embraces an absolute concept (inasmuch as the product is either sterile or not), the microbiological requirements for all other groups of pharmaceuticals, cosmetics and toiletries are by comparison much more flexible and generally intended to be applied with a degree of skilled interpretation. Additionally, the requirements are in many cases based upon guidelines and 'in-house' limits rather than on official standards. Such guidelines have evolved and have been refined during the last 30 years, based upon the information gathered from several independent surveys and from a considerable number of isolated case reports in the literature. The results from these surveys are well worth examining in detail since they provide the background to the rationale for the guidelines which subsequently developed.

15.3 Extent of microbial contamination

15.3.1 *Pharmaceuticals*

The nature and extent of the problem of microbiological contamination in pharmaceutical products was first fully appreciated by Kallings *et al.* (1966), following a survey for the Swedish National Board of Health of 600 unopened marketed drugs. A wide range of products was found to contain a variety of bacterial and fungal contaminants, including *Bacillus subtilis, Staphylococcus aureus* and other species, *Salmonella* spp. and coliforms, *P. aeruginosa, Alkaligenes* spp. and *Candida* spp.; in some cases, high microbial counts were found, in numbers exceeding 10^5 colony-forming units per gram (cfu g^{-1}) or per millilitre (cfu ml^{-1}).

Recognizing this to be an international problem, Kallings *et al.* proposed the introduction of microbial limits for non-sterile medicines and specifically an upper acceptable limit of 100 bacteria g^{-1} or ml^{-1}. In effect a requirement for a total viable count, as used today, had been recommended.

Later in 1971 the Public Health Laboratory Service reported on a survey of 1220 medicines in use in hospital wards in England and Wales (Anon. 1971a). Concern was expressed at the number of samples found to be heavily contaminated (18% contained in excess of 10^4 cfu g^{-1} or cfu ml^{-1}, indicating probable multiplication within the products. Moreover, the isolation of *P. aeruginosa* from 2.7% of samples (mainly in peppermint water, alkaline mixtures flavoured with peppermint water and aqueous topical cream) was viewed with particular concern. In the same year the Pharmaceutical Society of Great Britain reported on a similar but independent investigation into contamination in both commercially made and hospital-prepared products (Anon. 1971b). Raw materials of natural origin and finished products containing those materials were often found to be heavily contaminated, as were various grades of pharmaceutical waters and aqueous products. Both reports recommended proposals as to how the overall microbial quality of medicines might be improved in practice and urged that pharmaceutical requirements should include tests for the absence of harmful organisms. However, the notion of imposing microbial limit tests, as recommended by the Swedish authorities, was strongly resisted, although it was recognized that such limits had a place in individual manufacturer's in-process controls.

The presence of *P. aeruginosa* in pharmaceuticals continued to cause concern in subsequent years, as it was isolated from a wide range of hospital manufactured products. In one survey of 499 unused products, it was isolated from 9.2% of samples, mainly aqueous based creams and disinfectants (Baird *et al.* 1976). A strong correlation was shown to exist between medicament strains of *P. aeruginosa* and environmental strains isolated from the corresponding manufacturing facilities, suggesting that medicament strains had originated from environmental sources.

Further isolated case reports on contaminated pharmaceuticals continued to be published during the 1970s. Together these reports emphasized the need for official guidance on both what constituted an acceptable level of contamination in a non-sterile pharmaceutical product and which groups of organisms should be excluded by virtue of their potential pathogenicity.

15.3.2 Cosmetics and toiletries

The development of microbiological guidelines for cosmetic products and toiletries followed a similar path. A number of surveys over the decade 1965–1975 reported a varying incidence of contamination in unused cosmetics, ranging from 2.5% to 43% (Heiss 1967, Wolven and Levenstein 1969, Dunnigan and Evans 1970, Wolven and Levenstein 1972, Myers and Pasutto 1973, Jarvis *et al.* 1974, Anon. 1974, Baird 1977). Such wide variations in results as shown in Table 15.1 could be explained by a number of factors. Certain products, notably aqueous preparations, were known to be more

Table 15.1 Reported incidence of contamination in unused cosmetic products

Reference	Number of samples	Amount contaminated (%)	Product types
Heiss (1967)	60	43.3	Cream, cleanser, tonics
Wolven and Levenstein (1969)	250	24.4	Cream, cleanser, lotion, eye
Dunnigan and Evans (1970)	169	19.5	Cream, cleanser, lotions, baby
Wolven and Levenstein (1972)	223	3.6	Creams, cleansers, lotion, eye, baby, hair, bath
Myers and Pasutto (1973)	165	12.1	Cream, eye, hair, lipstick, powder
Anon. (1974)	2680	2.5	Cream, lotion, baby, dental, shaving, perfume, deodorant
Baird (1977)	147	32.7	Cream, cleanser, lotion, eye, baby, dental, hair

Copyright: *Econ. Microbiol.* Baird, R. M., 1981, **6** 404.

susceptible to contamination than others; hence the initial selection of samples may well have significantly affected the results reported. In some cases the surveys were concerned with only one product type, e.g. cosmetics for the eye (Wilson *et al.* 1971). Additionally, sampling and culture methods varied considerably between surveys. Higher contamination rates were found in those using an enrichment, rather than a direct culture technique, indicating the importance of incorporating a resuscitation step in testing such products. Furthermore, neutralization techniques used to inactivate antimicrobial agents varied from survey to survey, or indeed were overlooked in some cases.

More recent surveys on the incidence of contamination in cosmetic products have suggested that there has been an improvement in their microbial quality since the early 1970s. In a survey of 4499 unused cosmetics in the EEC, Greenwood and Hooper (1982) reported that only 5% contained greater than 10^3 cfu/g. Likewise Van Doorne (1992) found that 90% of 2,430 unused cosmetics contained less than 10^3 cfu g^{-1}.

Overall, these findings indicated, however, that, under the existing manufacturing conditions and constraints at that time, some form of contamination in the final product appeared to be inevitable. In particular, static populations containing low levels of aerobic spore bearers and Gram-positive cocci were regarded as unavoidable contaminants. However, the presence of Gram-negative contaminants was, regarded as undesirable, since these were usually present as dynamic populations with minimum nutritional requirements and could therefore multiply to high numbers. Furthermore, it was recognized that cosmetics and toiletry products were used in a variety of situations, and some could find their way into hospitals. Here, their microbial flora was of particular interest, since it was known that handcreams, handlotions, detergents and shampoos had been responsible for outbreaks of hospital infection on a number of occasions in the past (Morse *et al.* 1967, Victorin 1967, Cooke *et al.* 1970).

15.4 Current standards and guidelines

15.4.1 *Pharmaceuticals*

Earlier studies had thus identified two microbiological requirements, namely a total viable count and the absence of specified pathogens, which should clearly be incorporated into any proposed pharmacopoeial specifications. These two requirements subsequently formed the basis of specifications in use today, although they may not necessarily have been used together. Interestingly, the European countries, other than the UK, have tended to follow the Swedish initiative and selected a specification based upon a total viable count (TVC) of usually not more than 10^2 cfu g^{-1} or cfu ml^{-1}. In contrast, the *BP* (Anon. 1993a) and *United States Pharmacopoeia* (*USP*), (Anon. 1990a) until recently have resisted pressure to introduce total viable count limits but have favoured a specification requiring the absence of one or more indicator organisms in named products. The list of unacceptable organisms has increased over the years; the most recent *BP* requires the absence of *Escherichia coli*, Salmonella spp., *P. aeruginosa*, Enterobacteriaceae and certain other Gram-negative bacteria, *Staph. aureus* and pathogenic Clostridia in named products (Anon. 1993a). Appendix XVIB of the *BP* has also included a semi-quantitative test for *Clostridia perfringens* for those products where the level of this species is a criterion of quality. The range of material thus affected has remained small and selection of such products has been based upon a previously identified hazard, e.g. the absence of *Salmonella* spp. in carmine and pancreatin powders.

In the most recent *BP* reference is also made to additional test methods for the absence of certain specified micro-organisms and for a total viable aerobic count, as described in the *European Pharmacopoeia* (*EP* or *Ph. Eur.*) (Anon. 1986), since these tests may be invoked in certain European countries at the discretion of national pharmacopoeial authorities. Although not required as such by the *BP*, they are included for reference and information of users of the *BP*. It should be noted, however, that the *EP* test now takes precedence over any national pharmacopoeial test – see the Postscript at the end of this chapter.

The most recent edition of the *USP* (Anon. 1995a) requires not only the absence of specific microbial contaminants but also a microbial count limit for certain monographs.

Thus from the above discussion it can be seen that there has been a general move towards harmonization of the differing test requirements in recent years. Both official and unofficial microbial specifications applied to raw materials and finished products manufactured in the UK are summarized in Table 15.2.

In isolation, neither the total viable count nor specific tests are totally satisfactory. In considering the total viable count, several sampling factors must be taken into account: how the sample is taken, the sample size, and the point in time when the sample is taken, for viable counts may be dynamic populations. The handling of the sample itself can influence the microbial population; the time elapsing between taking and testing the sample, as well as the storage conditions under which it is held, may bring about changes in microbial bioburden levels. Furthermore, if microbial cells carry sublethal injuries, perhaps following exposure to preservatives, these may require specific resuscitation conditions before such injuries can be repaired and cell growth can again take place. Failure to provide these conditions may result in deceptively low total viable counts. Care is also required in interpreting numerical data especially when the sensitivity of the method may be questioned. An obvious difficulty here would be the problem posed by the use and interpretation of a strict limit test; in microbiological terms the difference between a total viable count of 90 cfu g^{-1} and 110 cfu g^{-1} may be of dubious significance; yet strictly speaking, these fall either side of a limit test of 10^2 cfu g^{-1}.

A more realistic approach involves the use of a range of limits to serve as acceptable, warning and action limits for a given course to be taken. For example counts of 10^2–10^3 cfu g^{-1} or cfu ml^{-1} would be regarded as within acceptable limits, 10^3–10^4 cfu g^{-1} or ml^{-1} as warning limits and counts exceeding 10^4 cfu g^{-1} or cfu ml^{-1} as action limits. Such an approach has been adopted by quality controllers in the health service (Baird 1985a, Davison 1990).

Another clear shortcoming with this test method is that no distinction is made between innocuous and harmful organisms. A similar argument may also be applied, however, to the alternative test method for absence of specified organisms. For example, although *P.aeruginosa* is regarded as a harmful organism, the acceptability or otherwise of *Pseudomonas* spp. is not mentioned as such. There is now considerable reported evidence in the literature that the presence of other opportunist Gram-negative organisms in pharmaceutical products should be regarded as undesirable (Bassett 1971, Parker 1972, Baird 1985b). This fact is recognized to some extent in the current *BP* requirement for the absence of *Enterobacteriacea* and certain other Gram-negative bacteria in specified pharmaceuticals; however, the prescribed method may well fail to detect free-living organisms such as pseudomonads which grow preferentially at a lower temperature (30–32°C, compared with the recommended 35–37°C).

Assessing the suitability of a product for use is not always as straightforward as it might seem. In addition to the identity of the contaminant, account must be taken of the

Table 15.2 Microbial specifications for raw materials and pharmaceuticals manufactured in the UK

Product	Absence of	Total viable count (cfu g^{-1} or cfu ml^{-1})	Specification
[a] Acacia, agar, alginic acid, aluminium hydroxide, digitalis, gelatin, pancreatin, starch (various), sterculia, tragacanth	Enterobacteriaceae/other Gram-negative bacteria *Escherichia coli* *Pseudomonas aeruginosa* *Salmonella* spp. *Staphylococcus aureus* *Clostridia*	Not applicable	*British Pharmacopoeia*
Raw materials and finished products	*Escherichia coli* *Pseudomonas aeruginosa* *Salmonella* spp. *Staphylococcus aureus* Other contaminants[b]	'In-house' limits (unpublished)	Licensed manufactured products
Raw materials and finished oral and topical products	*Clostridia* *Escherichia coli* Enterobacteriaceae Pseudomonads Salmonella *Staph. aureus*	10^1–10^{3c}	Hospital-manufactured products

[a] Specification varies, see individual monograph.
[b] Specification varies according to nature of product and perceived risk of contamination.
[c] Specification varies, see Table 15.4.

number of organisms present, the use to which the product may be put, i.e. the route by which it is to be given, and the groups of patients likely to receive the product. Thus, in formulating any microbiological standards, account should be taken of the most sensitive patient, accepting that this may produce a product of a higher standard than that required for other patients.

15.4.2 Cosmetics and toiletries

As with pharmaceutical products, general awareness of the problems of microbial contamination in cosmetics and toiletry products has increased during the past three decades; in particular the potential risk to health has been more fully appreciated. Recognizing this to be a universal problem and acknowledging the need for a unified approach, manufacturers in the UK have co-operated with one another in recent years, sharing their knowledge and experience, with the aim of improving the microbiological standards of products made by the industry. This has been achieved by agreement, rather than by legislation, under the direction of the Cosmetics, Toiletries and Perfumery Association (CTPA), founded upon a set of guidelines to be observed by member manufacturers. The current guidelines (Anon. 1990b) are based principally upon the acceptance of recommended microbiological limits, i.e. a total viable count; these guidelines are under revision at this time (Anon. 1996).

In applying these limits to the product in its final container at the time of manufacture, it is assumed that, after testing, the number of micro-organisms will subsequently either remain constant or decline. By the same token it is recognized that such limits are meaningless if contaminants are able to multiply in the product after testing.

Products intended for direct use around the eye and those recommended for use on babies are seen as critical products, for which the total viable count (i.e. aerobic bacteria, yeasts and moulds) should be less than 100 cfu g^{-1} or cfu ml^{-1}; for other products the total viable count should be less than 1,000 cfu g^{-1} or cfu ml^{-1}. In order to comply with these requirements, the CTPA recommend that in-house warning limits should be set at levels ten times lower, i.e. 10 and 100 cfu g^{-1} or ml^{-1} respectively. Furthermore, there is an additional requirement that, whenever colonies are observed on plate counts, the presence of harmful micro-organisms should be excluded. Whilst recognizing that definitions of harmful organisms may differ (see chapter 2, section 2.2), suitable test methods for the detection of obviously undesirable micro-organisms are included for *Clostridia* spp., *P. aeruginosa*, *S. aureus* and *Candida albicans*. Particular attention is also drawn to the significance of Gram-negative bacilli and especially *P. aeruginosa* in eye and baby products. Any species of *Pseudomonas* is considered to be an opportunist pathogen.

In applying such limits, the CTPA utilizes a three-class attribute plan, taking five randomly selected samples/batch or alternatively one random sample/batch of five consecutively produced batches. Resulting TVCs will then fall into one of three categories: an unacceptable category with a count higher than the upper limit known as *M* (100 cfu g^{-1} or ml^{-1} for baby or eye products, 1000 cfu g^{-1} or ml^{-1} for all other products); a hold and retest category with a count between *M* and a lower predetermined value known as *m* (set experimentally from testing 100 sample units from 20 batches) and usually set at a value ten times smaller than *M* but above background contamination levels; or an acceptable category with a count less than *m*. Table 15.3 summarizes the limits for both general and critical care products and details the maximum number of defective sample units allowed (c). According to the category into which the product falls, a recommended course of action is then set out (release, reject, hold and retest the original and additional samples). In all instances the TVC should not include any detectable harmful organisms.

As mentioned in chapter 2 and as discussed previously by Spooner (1977), discussions are currently under way within the European Cosmetics Trade Association (COLIPA) for European microbial purity guidelines for certain cosmetic products which are liable to present a risk to health. The proposed limits are given in Table 15.4. There are no legal standards in Europe at present but there is growing pressure in certain quarters for this to be changed. Elsewhere there is resistance to such a move, the emphasis being placed upon compliance with accepted guidelines, drawn up by industry, rather than with imposed, legally enforceable standards. A recent European directive 93/35/EEC (Anon. 1993b) has provided statutory microbiological requirements governing the production of cosmetic products; the manufacturer must retain information not only on the micro-biological specifications of the raw materials and the finished product but also on the microbiological control criteria of the cosmetic product, thus ensuring compliance with the requirement that the product must not cause damage to human health when applied under normal or reasonably foreseeable conditions of use.

A similar situation exists in the USA, where there are no statutory limits, although the Federal Food Drug and Cosmetic Act of 1938 as amended prohibits the adulteration of such products from interstate commerce. The presence of pathogens, or opportunist pathogens, would clearly constitute adulteration by virtue of the potential danger to health. As

Table 15.3 CTPA microbiological limits for cosmetic and toiletry products (Anon. 1990b)

Product	Upper limit[a] M	Lower limit[a] m	C
General	1000	100	2
Baby/eye	100	10	0

[a] cfu g^{-1} or ml^{-1}.
[b] *m* provides an early-warning system and should be 10 times smaller than *M*.
C, number defective sample units allowed.

Table 15.4 COLIPA-CTPA proposed limits for microbiological purity of cosmetics and toiletries (Anon. 1994)

Category	Use	TVC (cfu g^{-1} or ml^{-1})	
		Aerobic bacteria	Yeasts and moulds
1	Baby/eye products	$\not> 10^{2a}$	$\not> 10^2$
2	General products	$\not> 10^3$	$\not> 10^2$
3	Products containing natural raw materials[b]	$\not> 10^4$	$\not> 10^2$

[a] 10^2 interpreted as a maximum limit of acceptance of 5×10^2, and so on.
[b] For which the accepted microbial contamination of the raw material exceeds 10^3 cfu g^{-1} or ml^{-1}.

with the CTPA, microbiological guidelines have been drawn up by the Cosmetic, Toiletries and Fragrance Association (CTFA) and used successfully for several years now in the USA (Anon. 1993c, Tennebaum 1977). These too have been based upon recommended numerical limits for different categories of products, as summarized in Table 15.5. Additionally, there is a requirement that no product should have a microbial content recognized as harmful to the user. Although unspecified as such, methods are given for the examination of *S. aureus*, *E. coli* and *P. aeruginosa*.

15.4.3   In-house guidelines

Apart from the previously mentioned official standards and adopted guidelines, many manufacturers of both pharmaceutical preparations and cosmetic and toiletry products have over the years developed their own in-house microbiological guidelines and limits. Based upon detailed knowledge of plant history and full product records, these have provided an invaluable and practical working specification of what can be achieved routinely in individual units. Such specifications may well incorporate more rigorous microbial limits than the manufacturers' registered limits and frequently are based upon a more flexible approach with a range of limits to define normal and action levels for a given product. Clearly, these will vary not only from manufacturer to manufacturer, but also within a given manufacturing unit from product to product, according to the microbial load in the raw materials, the degree of processing during manufacture, the stringency of environmental controls, the inherent susceptibility of the product to microbial contamination and the effectiveness of any preservatives incorporated into the final product.

   One recent development in the pharmaceutical industry has involved the setting up of a UK group, known as the Pharmaceutical Microbiology Interest Group (PharMIG), to provide a forum where microbiologists from the industry could share information and problems. This group has attracted a considerable number of members and in its short history has already begun to tackle a number of microbiological issues, including the question of microbiological standards for non-sterile pharmaceuticals.

   A similar approach of drawing up in-house microbiological guidelines has been adopted over a number of years by pharmacists concerned with the manufacture of medicines in hospitals, and these have subsequently been adopted nationally (Baird 1985b, Davison 1990). In the past the manufacture of these products was poorly controlled and the microbiological quality and acceptability of finished products was therefore questionable. In particular, *P. aeruginosa* was a notorious contaminant in a

Table 15.5   CTFA microbiological limits for cosmetic and toiletry products (Anon. 1993c)

Product	Microbiological limit[a]
General	⋫ 1000[b]
Eye, baby	⋫ 500[b]

[a] cfu g^{-1} or ml^{-1}
[b] In addition to microbiological limits, no product to have a microbial content recognized as harmful to the user as recovered by standard plate count procedures.

range of products, including disinfectants, antiseptics, lotions, creams, alkaline mixtures and solutions (Baird *et al.* 1976). Thus, in drawing up microbiological specifications for use nationally within the service, quality control pharmacists adopted a two-pronged approach which combined both a requirement for an acceptable total viable count and a requirement for the absence of pathogenic or potentially pathogenic organisms, such as *S. aureus*, *Salmonella* spp., *E. coli*, Clostridia as well as Gram-negative opportunists bacteria, such as pseudomonads and Enterobacteriaceae. Specifications varied according to the type of product and its use. For example a distinction was made between raw materials and finished products (topical or oral). The specification also included a range of total viable counts encompassing an acceptable and maximum contamination level. The requirements are summarized in Table 15.6. For a more detailed discussion of testing methods for determining the total viable count, see chapter 3, section 3.6.

In comparing the development of both official standards and in-house guidelines in recent years, it is apparent that the widespread acceptance of such guidelines by industry and hospitals has to some extent influenced their subsequent inclusion in official standards.

15.5 Effect of implementing guidelines

The microbiological quality of hospital manufactured pharmaceuticals in this country has improved considerably in the past few years. There are a number of contributor factors including the upgrading of manufacturing facilities, the implementation of good manufacturing practice (GMP) (Anon. 1983, Anon. 1989), the impact of the Medicines Inspectorate in improving standards generally, and increased staff awareness of the problems of microbial contamination through education and training. Doubtless the acceptance and implementation of agreed national microbiological guidelines have also been instrumental in raising the standards by providing a specification against which the quality of a given product could be measured. Two surveys of hospital-produced pharmaceuticals have shown a considerable improvement in the overall quality of products,

Table 15.6 Microbial limits for raw materials and non-sterile categories of pharmaceuticals for hospital use

Criteria	Raw materials	Topical preparations	Oral preparations
TVC per g or ml			
Acceptable	10^2	10^1	10^2
Maximum	10^3	10^2	10^3
Absence in 1 g or 1 ml	*Staph. aureus*	*Staph. aureus*	*E. coli*
	Enterobacteriaceae	Enterobacteriaceae	Enterobacteriaceae[a]
	Pseudomonads	Pseudomonads	Pseudomonads[a]
	Clostridia		
Absence in 10 g or 10 ml	*Salmonella*		*Salmonella*

[a] Denotes upper limit of 10 per ml or g.
Copyright: Davison, A. L. (1990) *Guide to microbiological control in pharmaceuticals*, Ellis Horwood, Chichester, p. 363.

following the introduction of a comprehensive microbiological monitoring programme (Baird 1985a, b). In both cases a significant reduction in the general incidence of con- tamination was reported (mainly due to aerobic spore bearers and Gram-positive cocci), combined with a marked reduction in contamination levels due to Gram-negative oppor- tunist pathogens (Tables 15.7 and 15.8). Although unreported as such, it is probable that similar improvements have been made in the microbiological quality of commercially prepared products manufactured in the UK during the same time, as a result of the implementation of the previously discussed microbiological guidelines and limits. In Sweden the general microbiological quality of non-sterile pharmaceutical products is known to have improved considerably since the introduction of limit tests by the Swedish Board of Health in 1966 (Ringertz and Ringertz 1982).

15.6   Future trends

In drawing up and using microbiological specifications for pharmaceuticals, cosmetic and toiletry products, much can be learned from experience gained in other microbiological fields, and particularly from food microbiology. The microbiological analysis of all these commodities has, of course, a common objective, i.e. to assess the risk of microbial spoilage and the risk to consumer health associated with the use of a given product. Microbiological limits have been used successfully for many years in the food industry to achieve this very objective. Invaluable information can be obtained not only from study- ing how such limits have evolved but also by examining some of the wider issues, such as the sampling plans adopted, the recovery and testing methods used, the selection of marker organisms and the criteria for quality.

The three-class attribute sampling plan has found increasing favour in recent years in the food industry, the cosmetic industry (Anon. 1990b) and to a certain extent in the pharmaceutical industry (see chapter 3, section 3.3). It provides a greater degree of confidence in the quality of the product and it also forewarns of any potential problems at

Table 15.7   Contamination in 13,371 non-sterile products made in a hospital pharmacy: 1975–1983

Year	Number tested	Amount (%) contaminated with the following		
		Aerobic spore bearers	Gram-positive cocci	Gram-negative rods
1975	1 390	7.41	0.65	0.22
1976	1 791	7.93	0.84	0.73
1977	1 569	4.91	0.57	0.06
1978	1 583	3.79	1.96	0.13
1979	1 474	6.92	1.56	0.54
1980	1 376	5.81	2.95	0.22
1981	1 415	4.66	2.69	0.14
1982	1 486	3.84	1.35	0
1983	1 287	3.57	0.78	0
Total	13 371	5.48	1.46	0.24

Copyright: *Pharm. J.* (1985) **234** 54.

Table 15.8 Incidence of contamination in 4,369 non-sterile products made in nine hospital pharmacies: 1982–1984

Hospital	Amount (%) of contamination		
	1982	1983	1984
1	66.6	13.8	13.4
2	13.8	3.5	1.7
3	NT	7.5	8.2
4	25.0	13.9	7.5
5	18.0	4.2	1.6
6	NT	22.7	5.8
7	NT	15.9	10.9
8	34.2	12.1	7.1
9	47.0	25.7	20.3
Average	30.9	10.2	8.0

NT, not tested.
Copyright: *J. Clin. Hosp. Pharm.* (1985) **10** 95.

an earlier stage. The scheme incorporates so-called tolerances and permits a small population of samples to show slight deficiencies, i.e. not exceeding a specified predetermined count by a factor over 10. The plan acknowledges that samples will fall into one of three classes – fully acceptable, marginally acceptable or unacceptable – and describes the course of action to be taken accordingly, i.e. release of the batch, retest batch, hold and retest batch, treat and retest batch, or reject the batch. Clearly the use of the three-class sampling plan could be extended to other types of product, besides cosmetics and toiletries.

In drawing up any microbiological specifications, the major issues will obviously be of prime concern and must be addressed first. The presence of any micro-organisms considered to be pathogenic is clearly one such issue. Traditionally the designation of these organisms has been determined in part by demonstrating that a hazard has previously existed. For example, there were no microbiological specifications for thyroid until 1966 when Kallings *et al.* demonstrated, firstly, the susceptibility of this product to contamination by *Salmonella* spp. and, secondly, the risk to health from using such a product. Similarly, a wealth of documentary evidence over the past 40 years has conclusively demonstrated the hazards of *P. aeruginosa* contamination in a wide range of pharmaceutical products (Bassett 1971, Parker 1972, Ringertz and Ringertz 1982). Consequently, some of these products must now be manufactured as sterile products whilst others must be shown to be free from *P. aeruginosa* contamination. In both instances there is an undisputable course of action to be taken should such contamination be detected.

Less obvious is the course of action required when other micro-organisms are isolated, as in the case of Gram-negative opportunist pathogens which are frequently found in practice. Widely distributed in the environment of both hospitals and factories, these water-borne organisms are characterized by having minimal nutritional requirements, thus enabling them to survive and multiply in some unlikely situations including pharmaceutical, cosmetics and toiletry products. Here, their presence in such products may be found disturbing, depending on the products' intended use. Past experience indicates that the microbiological content of baby and eye products must be regarded as critical and contamination with such opportunist pathogens is clearly unacceptable. Other cosmetics and toiletries, notably hand-creams and hand-lotions may make their way into hospitals and here their microbiological flora may be of more concern than in a domestic environment (Morse *et al.* 1967).

In a similar way Gram-positive rods and cocci have traditionally been regarded as fairly innocuous but inevitable contaminants of pharmaceutical and toiletry products. Widely distributed in the manufacturing environment and found in many raw materials, their presence in the finished product has been largely accepted (Baird 1985a, b). A wide variety of *Bacillus* spp. has been isolated from pharmaceutical products (Willense-Collinet *et al.* 1981, Gil *et al.* 1986, de la Rosa *et al.* 1993); *Staph. epidemidis* has been a common isolate too (Devleeschouwer and Dony 1979, Devleeschouwer *et al.* 1980, Garcia Arribas *et al.* 1983, Van Doorne and Boer 1987, de la Rosa *et al.* 1993). Recent concern has been expressed, however, regarding the level of antibiotic resistance exhibited by some of these strains (de la Rosa *et al.* 1993).

Profound changes have of course been witnessed in the past 40 years or so in hospital practice, particularly in the type of patients who are now being treated there. The widespread use of chemotherapeutic, antibiotic, immunosuppressant or cytotoxic agents on susceptible individuals, such as neonates, the elderly and others debilitated by intensive and advanced surgery, has brought about significant changes in the types and numbers of infections seen in hospitals today, and to a lesser extent in the general community. Opportunist Gram-negative pathogens have been held responsible for an increasing number of infections in such patients. Some of these infections are known to have originated from contaminated pharmaceutical, cosmetic and toiletry products. Incidents of infection associated with contamination in pharmaceuticals, cosmetics and toiletries were reviewed in detail in chapter 2 from which it can be seen that many of these incidents have been characterized by the involvement of a previously unappreciated risk in a group of susceptible and debilitated patients. Infections due to *Pseudomonas* spp. have been the most notorious group in the past, but other organisms such as *Acinetobacter* spp., *Serratia* spp., *Aeromonas* spp., *Alkaligenes* spp. and *Enterobacter* spp. have also featured increasingly often (Ramphal and Kluge 1979). At the same time, it should be remembered that reported cases represent a mere fraction of incidents occurring in practice, many of which remain undetected for a variety of reasons.

Thus, in returning to the vexed question of which organisms may be regarded as acceptable contaminants in pharmaceutical, cosmetic and toiletry products, one possible approach is to assess the potential risk to health by evaluating all known contributory factors. These include the type and number of contaminating organisms, the inherent susceptibility of the product to contamination, the efficacy of any incorporated preservative system, the route by which the product is to be used or given, whether the product is intended for single- or multiple-dose application by individual or multiple users, and the likely health status of any user of the product. In some cases, these factors may simply not be known; in other cases, they may be uncontrollable, as the product may be used in a variety of situations. In such cases, there is an argument for providing products of an acceptable microbiological quality for the most susceptible patient concerned.

Microbiological specifications for all pharmaceutical, cosmetic and toiletry products should reflect current information and knowledge. From time to time, they will require revision. Previous experience indicates that, as time goes by, specifications are likely to become tighter and more rigid. In the past three decades, there has been a dramatic improvement in our understanding of why contamination occurs, how it can be controlled and why it is necessary to do so.

Complete control can, of course, be easily achieved by packaging all finished products in single-dose units and subjecting them to appropriate sterilizing processes. Heavy financial penalties are naturally associated with this course of action, not only for manufacturers, but also for health services, patients and consumers. Common sense indicates that satisfactory control can be achieved by other means and without the requirement for specific legislation.

15.7 Postscript

The EP has recently published microbial quality criteria for non-sterile pharmaceutical products (Anon. 1995b). These vary according to the intended use of the product: topical or use in the respiratory tract; oral or rectal administration; those used as herbal remedies. In all instances the quality criteria include a total viable count limit for both bacteria and fungi and a required absence of specified organisms.

References

ALLEN, H. F. (1959) Aseptic technique in ophthalmology. *Trans. Am. Ophthalmol. Soc.* **57** 377–472.

ANON. (1932) *British Pharmacopoeia*, 6th edn, Constable, London, Appendix 16, p. 632.

(1966) *British Pharmaceutical Codex 1963*. Pharmaceutical Press, London, p. 77.

(1968a) *British Pharmacopoeia*. Pharmaceutical Press, London, p. 409.

(1968b) Medicines Act. HMSO, London.

(1971a) Microbial contamination of medicines administered to hospital patients. *Pharm. J.* **207** 96–99.

(1971b) Microbial contamination in pharmaceuticals for oral and topical use. *Pharm. J.* **207** 400–402.

(1974) Science Conference Committee Reports (1975). *CTFA Cosmet. J.* **7** 3.

(1983) *The guide to good pharmaceutical manufacturing practice*. HMSO, London.

(1986) *European Pharmacopoeia*. Second Edition. (see 1983 V.2.1.8.1-5 and 1991 V.2.1.8.9-10) Maisonneuve SA, Sainte-Ruffine, France.

(1989) *The rules governing medicinal products in the European Community Vol. IV. Guide to good manufacturing practice for medicinal products*. HMSO, London.

(1990a) *United States Pharmacopoeia* XXII. US Pharmacopoeial Convention, Rockville, Maryland pp. 1684–1685.

(1990b) *Microbial quality management limits and guidelines*. Cosmetic, Toiletry and Perfumery Association, London.

(1993a) *British Pharmacopoeia* Vol II. HMSO, London. Appendix XVIB A184–A190.

(1993b) Council Directive 93/35/EEC. Sixth Amendment Directive 76/768/EEC. *Off. J. Eur. Com.* **L151** 32–36.

(1993c) *Microbiological limits for cosmetics and toiletries*. Cosmetic, Toiletry and Fragrance Association Microbiology Guidelines.

(1994) *Cosmetic good manufacturing practices*. COLIPA, Brussels.

(1995a) *United States Pharmacopoeia*. XXIII US Pharmacopoeial Convention, Rockville, Maryland, p. 1939.

(1995b) *European Pharmacopoeia* (Second Edition) VIII 15.1–15.2.

(1996) Cosmetic, Toiletry and Perfumery Association, London. In Press.

BAIRD, R. M. (1977) Microbial contamination of cosmetic products. *J. Soc. Cosmet. Chem.* **28** 17–20.

(1985a) Microbial contamination of pharmaceutical products made in a hospital pharmacy: a nine year survey. *Pharm. J.* **231** 54–55.

(1985b) Microbial contamination of non-sterile pharmaceutical products made in hospitals in the North East Thames Regional Health Authority. *J. Clin. Hosp. Pharm.* **10** 95–100.

BAIRD, R. M., BROWN, W. R. L., and SHOOTER, R. A. (1976) *Pseudomonas aeruginosa* infections associated with the use of contaminated medicaments. *Br. Med. J.* **2** 349–350.

BASSETT, D. C. J. (1971) Causes and prevention of sepsis due to Gram-negative bacteria: common source outbreaks. *Proc. Soc. Med.* **64** 980–986.

COOKE, E. M., SHOOTER, R. A., O'FARRELL, S. M., and MARTIN, D. R. (1970) Faecal carriage of *Pseudomonas aeruginosa* by new-born babies. *Lancet* ii 1045–1046.

CROMPTON, D. O. (1962) Ophthalmic prescribing. *Australas. J. Pharm.* **43** 1020–1028.

DALE, J. K., NOOK, M. A., and BARBIERS, A. R. (1959) Effectiveness of preservatives in commercial ophthalmic preparations. *J. Am. Pharm. Assoc.* **20** 32–35.

DAVIES, D. J. C., and NORTON, D. A. (1975) Challenge test of antimicrobial agents. *Pharm. J.* **27** 383–384.

DAVISON, A. L. (1990) Microbial standards for pharmaceuticals. In: Denyer, S., and Baird, R. (eds), *Guide to microbiological control in pharmaceuticals*. Ellis Horwood, Chichester, pp. 356–365.

DE LA ROSA, M. DEL C., MOSSO, M. A., GARCIA, M. L., and PLAZA, C. (1993) Resistance to the antimicrobial agents of bacteria isolated from non-sterile pharmaceuticals. *J. Appl. Bact.* **74** 570–577.

DEVLEESCHOUWER, M. J., and DONY, J. (1979) Flore microbienne des medicaments. Données ećologiques et sensibilité aux antibiotiques. 1ª partie: formes magistrale seches. *J. Pharm. Belg.* **34** 189–203.

DEVLEESCHOUWERE, M. J., AKIN, A., and DONY, J. (1980) Flore microbienne des medicaments. Données écologiques et sensibilité aux antibiotiques. 2ª partie: formes magistrale liquide. *J. Pharm. Belg.* **35** 266–272.

DUNNIGAN, A. P., and EVANS, J. R. (1970) Report of a special survey: microbiological contamination of topical drugs and cosmetics. *Toilet. Goods Assoc. Cosmet. J.* **2** 39–41.

GARCIA ARRIBAS, M. L., DE LA ROSA, M. C., and MOSSO, M. A. (1983) Caracterisation et sensibilité aux antibactèriens des souches de coques Gram positif isolées a partir de medicaments solides administrées par voie orale. *J. Pharm. Belg.* **38** 140–146.

GIL, M. C., DE LA ROSA, M. C., MOSSO, M. A., and GARCIA ARRIBAS, M. L. (1986) Numerical taxonomy of *Bacillus* isolated from orally administered drugs. *J. Appl. Bact.* **61** 347–356.

GREENWOOD, M. H., and HOOPER, W. L. (1982) Survey report on the microbiological quality of cosmetics and toiletries on sale within the countries of the EEC.

HEISS, F. (1967) Keimgehalt von Korperpflegemitteln. *Fette Seifen, Anstrichm.* **69** 365–369.

JARVIS, B., REYNOLDS, A. J., RHODES, A. C., and ARMSTRONG, M. (1974) A survey of microbiological contamination in cosmetics and toiletries in the UK (1971). *J. Soc. Cosmet. Chem.* **25** 563–575.

KALLINGS, L. O., RINGERTZ, O., SILVERSTOLPE, L., and ERNERFELDT, F. (1966) Microbiological contamination of medical preparations. *Acta Pharm. Suec.* **3** 219–230.

MORSE, L. J., WILLIAMS, H. L., GREEN, F. P., ELDRIDGE, E. E., and ROTTA, J. R. (1967) Septicaemia due to *Klebsiella pneumoniae* originating from a handcream dispenser. *New Engl. J. Med.* **277** 472–473.

MYERS, G. E., and PASUTTO, F. M. (1973) Microbial contamination of cosmetics and toiletries. *Can. J. Pharm. Sci.* **8** 19–23.

NORTON, D. A., DAVIES, D. J. G., MEAKIN, B. J., and KEALL, A. (1974) Antimicrobial efficiency of contact lens solutions. *J. Pharm. Pharmacol.* **26** 841–846.

PARKER, M. T. (1972) The clinical significance of the presence of micro-organisms in pharmaceutical and cosmetic preparations. *J. Soc. Cosmet. Chem.* **23** 415–426.

RAMPHAL, R., and KLUGE, R. M. (1979) *Acinetobacter calcoaceticus* variety *anitratus*: an increasing nosocomial problem. *Am. J. Med. Sci.* **277** 57–66.

RICHARDS, R. M. E. (1972) Contact lenses and their solutions. *Pharm. J.* **208** 314–316.

RICHARDSON, N. E., DAVIES, D. J. G., MEAKIN, B. J., and NORTON, D. A. (1977) Loss of antibacterial preservatives from contact lens solutions during storage. *J. Pharm. Pharmacol.* **29** 717–722.

RINGERTZ, O., and RINGERTZ, S. (1982) The clinical significance of microbial contamination in pharmaceutical and allied products. *Adv. Pharm. Sci.* **1** 201–225.

SPOONER, D. F. (1977) Microbiological aspects of the EEC Cosmetics Directive. *Cosmet. Toilet.* **92** 42–51.

TENNEBAUM, S. (1977) Considerations leading to the development of microbial limit guidelines of the Cosmetics, Toiletries and Fragrance Association. *Am. Cosmet. Perfum.* **92** 79–83.

VAN DOORNE, H. (1992) Fundamental aspects of preservation of cosmetics and toiletries. *Parfum. Kosmetick.* **73** 84–92.

VAN DOORNE, H., and BOER, Y. (1987) Microbiological quality of hand-filled gelatin capsules. *Pharm. Weekblad* **122** 820–823.

VICTORIN, L. (1967) An epidemic of otitis in new borns due to infections with *Pseudomonas aeruginosa*. *Acta Paediatr. Scand.* **56** 344–348.

WILLENSE-COLLINET, M. F., TURNBULL, P. C. B., HOSPERG, G. T., and VAN OPPENRAAY, A. B. (1981) Computer-assisted method for identification of *Bacillus* species isolated from liquid antacids. *Appl. Env. Microbiol.* **41** 169–172.

WILSON, L. A., KEUHNE, J. W., HALL, W., and AHEARN, D. G. (1971) Microbial contamination in ocular cosmetics. *Am. J. Ophthalmol.* **71** 1298–1302.

WOLVEN, A., and LEVENSTEIN, I. (1969) Cosmetics – contaminated or not. *Toilet Goods Assoc. Cosmet. J.* **1** 34.

(1972) Microbiological examination of cosmetics. *Am. Cosmet. Perfum.* **87** 63–65.

Index